T0276155

CAMBRIDGE LIBRARY COLLECTION

Books of enduring scholarly value

Botany and Horticulture

Until the nineteenth century, the investigation of natural phenomena, plants and animals was considered either the preserve of elite scholars or a pastime for the leisured upper classes. As increasing academic rigour and systematisation was brought to the study of 'natural history', its sub-disciplines were adopted into university curricula, and learned societies (such as the Royal Horticultural Society, founded in 1804) were established to support research in these areas. A related development was strong enthusiasm for exotic garden plants, which resulted in plant collecting expeditions to every corner of the globe, sometimes with tragic consequences. This series includes accounts of some of those expeditions, detailed reference works on the flora of different regions, and practical advice for amateur and professional gardeners.

Rough Notes of a Journey through the Wilderness, from Trinidad to Pará, Brazil

Sir Henry Alexander Wickham (1846–1928) is remembered for his role in bringing the seeds of the rubber tree in 1876 from Brazil to the Royal Botanic Gardens at Kew, where seedlings were successfully cultivated and then sent to Asia for the establishment of commercial plantations. Wickham later styled his actions in collecting some 70,000 seeds as a tale of botanical smuggling, though at the time such action was not illegal. Skilled as a self-publicist, he enjoyed the great acclaim of the rubber industry as it burgeoned in British colonies abroad. This account, first published in 1872, is of Wickham's earlier travels in South America. The first part of the work traces his journey by river into the continent, recording his observations on rubber cultivation in Brazil. The second part describes his time among the indigenous peoples who lived on the Caribbean coast of Central America.

Cambridge University Press has long been a pioneer in the reissuing of out-of-print titles from its own backlist, producing digital reprints of books that are still sought after by scholars and students but could not be reprinted economically using traditional technology. The Cambridge Library Collection extends this activity to a wider range of books which are still of importance to researchers and professionals, either for the source material they contain, or as landmarks in the history of their academic discipline.

Drawing from the world-renowned collections in the Cambridge University Library and other partner libraries, and guided by the advice of experts in each subject area, Cambridge University Press is using state-of-the-art scanning machines in its own Printing House to capture the content of each book selected for inclusion. The files are processed to give a consistently clear, crisp image, and the books finished to the high quality standard for which the Press is recognised around the world. The latest print-on-demand technology ensures that the books will remain available indefinitely, and that orders for single or multiple copies can quickly be supplied.

The Cambridge Library Collection brings back to life books of enduring scholarly value (including out-of-copyright works originally issued by other publishers) across a wide range of disciplines in the humanities and social sciences and in science and technology.

Rough Notes of a Journey through the Wilderness, from Trinidad to Pará, Brazil

*By Way of the Great Cataracts
of the Orinoco, Atabapo, and Rio Negro*

HENRY ALEXANDER WICKHAM

CAMBRIDGE
UNIVERSITY PRESS

CAMBRIDGE
UNIVERSITY PRESS

University Printing House, Cambridge, CB2 8BS, United Kingdom

Published in the United States of America by Cambridge University Press, New York

Cambridge University Press is part of the University of Cambridge.

It furthers the University's mission by disseminating knowledge in the pursuit of
education, learning and research at the highest international levels of excellence.

www.cambridge.org
Information on this title: www.cambridge.org/9781108070041

This edition first published 1872
This digitally printed version 2014

ISBN 978-1-108-07004-1 Paperback

ROUGH NOTES

OF

AMERICAN TRAVEL.

CIRINGA TREE, (Indian Rubber)
Extracting Rubber.

ROUGH NOTES

OF A

JOURNEY THROUGH THE WILDERNESS,

FROM

TRINIDAD TO PARA, BRAZIL,

BY WAY OF THE

GREAT CATARACTS

OF THE

ORINOCO, ATABAPO, AND RIO NEGRO.

BY

HENRY ALEXANDER WICKHAM.

WITH ILLUSTRATIONS DRAWN ON THE SPOT BY THE
AUTHOR.

LONDON :
W. H. J. CARTER, 12, REGENT STREET, S.W.
———
1872.

TO

JAMES DE VISMES DRUMMOND HAY, C.B.,

H.B.M. CONSUL FOR VALPARAISO,

(LATE OF PARÁ,)

THESE ROUGH NOTES ARE DEDICATED,

IN REMEMBRANCE OF THE MANY KINDNESSES

BY WHICH

THE AUTHOR WAS INDEBTED FOR A PLEASANT ENDING

TO A SOMEWHAT ARDUOUS JOURNEY.

Santarem, Pará, 1871.

NOTICE.

THESE ROUGH NOTES OF AMERICAN TRAVEL, having been arranged and prepared for the Press without the personal supervision of the Author, may contain some errors, for which he cannot be responsible; but under the circumstances of his absence, all possible care has been given to the undertaking by those engaged in it.

CONTENTS.

PART I.

ROUGH NOTES of a JOURNEY THROUGH THE WILDERNESS, from Trinidad to Pará, Brazil, by way of the Great Cataracts of the Orinoco, Atabapo, and Rio Negro.

CHAPTER I.

CHAPTER II.

CHAPTER III.

CHAPTER IV.

CHAPTER V.

CHAPTER VI.

PART II.

A JOURNEY among the WOOLWA or SOUMOO INDIANS of Central America.

CHAPTER I.

CHAPTER II.

CHAPTER III.

CHAPTER VII.

CHAPTER VIII.

CHAPTER IX.

LIST OF ILLUSTRATIONS.

ROUGH NOTES

OF A

JOURNEY THROUGH THE WILDERNESS,

FROM TRINIDAD TO PARÁ, BRAZIL.

PART I.

ROUGH NOTES of a JOURNEY through the WILDERNESS, from Trinidad to Pará, Brazil, by way of the great Cataracts of the Orinoco, Atabapo, and Rio Negro.

CHAPTER I.

NEW YEAR'S EVE, 1869, we lay at St. Thomas's and witnessed a curious effect in the sky : a rainbow at night, caused by the moon-rays falling on a rain-cloud. On New Year's morning we steamed from the harbour, the shore of which and the neighbouring cays were still strewn with the hulks, wrecks, and other débris of the previous year's hurricane and earthquake, although now all seemed bathed in an atmosphere of undisturbed tranquillity. The Royal West India Mail ships in these latitudes are manned by blacks; it appeared quite natural to be again on the deck of the old "Tamar," watching the dusky forms of the crew as they lounged or romped about the forecastle, like so many monkeys.

B

Leaving the island of Santa Cruz on our starboard quarter, we sped along against a head wind. Some time after I had turned in, I was aroused by the report of the signal-gun, fired right over the port of my cabin : jumping into my slippers, I found we had arrived off St. Kitt's. A good many vessels lay in the roadstead ; and what appeared to be a town, slept on the lowland betwixt the water and the mountainous backbone of the island, over which rolled heavy manes of moon-lit cloud. The night was marked by a very beautiful, but here, not uncommon phenomenon; the top of each fleecy cloud was tinged with prismatic colours as it passed the disc of the moon. Next morning the sea-girt rock called Rodonda met our view; then Montserrat; and we coasted Antigua until we came to in the lovely cove-like entrance to the English harbour, with its crystalline water; there we landed mails. In the afternoon, looming suddenly through a mist of luke-warm rain, we saw the forest-covered coast of Guadaloupe, and I managed, as usual, to take a few rough sketches at any point of interest. The scenery was truly magnificent. Leaving at midnight, we proceeded to land mails at the Island of Dominica, from whence the gentle land-breeze wafted us that delightful fragrance peculiar to the tropical forests.

JANUARY 3RD.—Martinique by daybreak. We came to St. Lucia before noon, entering the beautiful harbour to remain the greater part of the day. When we were about to get under weigh, three masts, with long black spars, were visible, rounding

to face page 3.

SANTA LUCIA.

the promontory, and a large war-ship swept majestically across the narrow outlet: she proved to be the " Royal Alfred," bearing the pennant of the Admiral of the West India station. We were, of course, boarded for mails, papers, &c. What a difference there is in the appearance of the boat's crew from an English man-of-war, on a foreign station, to the sailors belonging to any other power!

St. Lucia was my first vision of the tropics on my way to Central America. I was then enchanted ; and though I have seen so much since the commencement of my travels, this isle still seems, above all places, to be transcendent in beauty ; the very rocks are robed in the deepest green. Strangely enough, we rounded the S.W. end again (where stand the giant Pitons), at sunset, the same hour as on my introductory visit : these peaks rise abruptly from the sea to a height of 4,000 feet.

On the 4th we lay all day at Grenada, taking in coal, which gave any of the passengers so inclined a good opportunity for a run ashore. The quaint, clean little town is very ancient and picturesque ; it is well situated in a fine bay, and, for the most part, on a small peninsula therein. In the morning we crossed the roughly-paved streets, ascending the side of the hill by a winding, but evenly cut, military road. The view from an old fort at the summit was worth the journey. In the afternoon we skirted the bay through thickets of mimosa and mangoe, interspersed with coco and other palms. I managed to have a swim ; but coming unexpectedly upon a shallow covered with sea-hedgehogs, I wounded my knee rather painfully.

We had on board a recruiter of the 4th West India Regiment, a very fine specimen of a West Indian soldier; his uniform excellently suited the exigencies of the climate, resembling that of the French Turcos. He might have passed for one of that dashing fraternity in Paris.

On reaching Trinidad I determined to deposit my baggage at the Custom-House by the water side, in order to be ready at the next opportunity for pushing on to the main.

I found that the steamer did not set out for Ciudad Bolivar before the 28th; and therefore, not desiring to remain stationary for such a length of time, I looked anxiously for a lancha or canoe, or anything in the shipping line that would answer my purpose, bound up the Orinoco.

9TH.—I took Rogers (a young Englishman who accompanied me) away behind the town across the country, to try his powers of locomotion and hardihood, knowing that a somewhat fatiguing walk lay before us. We gained the summit of a hill towards the north, where a most beautiful view of the flat coast and the northern mountains was spread beneath us. I think that any companion was duly impressed that travels such as we had entered upon would be no light pastime for either of us.

Mr. Budge, of Port of Spain, kindly placed his cool little cottage at my disposal. I was glad of this excuse to " clear out" of the Royal Hotel, as I found hotel life here, of necessity, expensive.

The town of Port of Spain, although not prepossessing from the water, is well laid out. The large number of East Indian coolies employed on the neighbouring cacao and sugar estates, give the streets something of an Eastern character: the town population is 25,000; and many more white faces are seen than in the other islands we touched at on our way.

Besides the conventional West Indian negro and the coolies, a sprinkling of Chinese is observable amongst the inhabitants; and I believe that the whole island population is some 90,000. I walked to St. Joseph's, the site of the old Spanish capital San José de Oruña; and, in company with Mr. Budge, visited the valley of Diego Martin.

On the 11th my patience was rewarded; a small vessel turned up bound for the required destination; so I at once engaged with its captain to convey Rogers and myself as far as Barrancas, and thence to Las Tablas, near the mouth of the Caroni tributary. We hoped to arrive in eight days.

Next day we left Port of Spain, and El Capitan ran the little craft up a creek to take in provisions. We encamped in a mangrove swamp, no very delectable situation; for no sooner had the sun disappeared, than we were beset by innumerable mosquitos and sand-flies. The latter insects, though of the minutest proportions, cause much irritation of the skin; and aided in their attacks by some full-grown trumpeting mosquitos, they rendered night hideous, and sleep quite out of the question. However, every disagreeable must come to an end, and so did ours at last: we struggled with difficulty out of the black mud, and, provisions on board, floated beyond the Chowan, as the detestable place is called.

Along the Trinidad coast, past the neighbourhood of the famous pitch-lake, the woods began to assume the primeval type. Cedras Point was rounded in the afternoon: the verdant green with which it is draped forms a peculiarly striking contrast to its own red colouring.

The complexion of the mighty Orinoco here tinges the water with yellow.

We passed the night at Cedras, swinging at anchor near the sandy shore.

13TH.—We left behind us some surf-worn sandstone rocks, covered with pelicans, and stood across the gulf, beyond the long reef designated "The Soldier"; coming upon the well-defined line where the greener water of the sea is borne back by the yellow tide of the Orinoco. There was little breeze, and so swelteringly hot in our open boat, it was determined to gain the delta of the Orinoco by the Pedernales channel. The sand-spits round the mouth were bright with scarlet ibis, egrets, &c. There are a few clearings just inside, and here we saw the Warraw or Guarrauno Indians (the tribe indigenous to these delta lands), in their

to face page 6.

SUN RISE, ORINOCO.

"curiaras," or canoes. As we advanced, channels opened out on every side, cutting the deep green mangrove woods into islands of various dimensions. It was evening, and as the sun dipped, macaws in pairs, and parrots in flocks, sociably chattering, wended their way to favourite roosts.

At dark the tide turned, and ran out so strongly, that we were obliged to make fast to the bushes, and wait for the flow.

14TH.—Now and then we encountered a curiara; but as soon as our boat was perceived by its Guarrauno freight, they made off, paddling as for dear life, vanishing up some friendly creek, amidst the dense thickets. This race has no doubt been badly treated by the Spaniards; kidnapped, and carried to work far away, not knowing whither, by armed forces. They are believed to be very numerous, but, owing to bad treatment, are very shy and difficult to approach. As well as the delta lands of the Orinoco, they inhabit the low coast between this and the Pomeroon, towards the Arrawaks of the Demerara; living generally far back in the swamps, where they lift simple palm-thatched lodges on high posts above the ooze. For their sustenance, they rely largely on a substance resembling sago, together with fish, which is abundant, and wild hog. In person they are short, and rather squat, probably from being so much in their canoes. In colour, they are darker than other American Indians I have seen, with the exception of the Moskitos of Central America. I was fortunate in obtaining a good subject for a sketch; the dusky individual showed great taciturnity (or indifference) during the time I kept him standing. These people occasionally make a trip in their curiaras to the neighbouring island of Trinidad, where they dispose of hammocks, tame birds, and other articles of Indian traffic. They also go to the British settlements on the Guayana coast, but for the most part keep within the labyrinth of creeks and channels connecting the main estuaries of the Orinoco, forming a watery maze, the secret of which they alone are in possession of. They are said also to possess the knowledge of an ointment that is obnoxious to mosquitos, which cease to torment them after they have anointed their bodies with the valuable charm. I think there really must be some ground for the idea, for I have seen the Indian fires shining through the night in swampy localities, where we, in an unprotected state, found it impossible to obtain the least respite from the pests.

As we advanced, the palm called the Manriche became more and more abundant. The Guarrauno make from it a farinaceous substance, which forms one of their staple supports.

15th.—Last evening, and in the morning, we were favoured with regular Orinoco squalls of wind and rain; later in the day, during our halt, I got out a strong line and some large hooks; we soon took five fine fish, which proved excellent eating— they were something like cat-fish. Presently we began to lose the mangrove, and to see less of the

MANRICHE PALM.

slender, graceful manac palm. In Brazil it is called the Assai, and it is there much esteemed for the refreshing beverage made from its fruit.

The mosquitos of the delta are truly terrible; they rise in clouds at night from the banks and the floating aquatic plants to prosecute their vengeance upon intruding man. Islands of these plants float up and down with the ebb and flow of the tide, filled with swarms of these plagues. The banks became higher and more habitable in aspect; but in the afternoon the wind so freshened against us, that we made fast to a cometure tree. As a lad named Pedro was about to perform this office, he nearly ran his head against a large mapanare snake, which was asleep on a bough. The men called out to me to shoot "very bad snake: he was accordingly riddled with a charge of No. 8 shot I happened to have in readiness for small birds. This snake is in the habit of lying thus coiled on a thick branch or bush over the water. I shot many in that way afterwards, and once or twice narrowly escaped brushing them down into the canoe in our passage underneath. Our little craft, about the size of a Margate lugger, was well manned; the crew were all excellent fellows in their way, although confirmed smugglers; indeed, the boat was afterwards confiscated by the authorities at Angostura. Benning, the master, a Trinidad creole, had been to England, and had seen the Crystal Palace, Thames Tunnel, and other objects of note; Paul, a brown man from Guadaloupe, chiefly steered;

Andreas, who cooked, was a Trinidad creole,—he had traversed the Spanish main ever since he was twelve years old, and had gone over that extraordinary bifurcation of the waters of the Orinoco with the Amazon, by means of the Cassiquiare, which runs from the upper Orinoco into the Rio Negro, or Guariana; lastly, Pedro, a mestizo (half-Indian), from the island of Margarita, off the north coast. The mosquitos continued to be most distressingly numerous.

16TH.—The banks of the channel reminded me of the lower part of Blewfields, in Central America, but water-birds and iguana were scarce here, perhaps owing to the deep water up to the banks. There was, however, an abundance of game, but we had not leisure to seek it on shore with gun and machete. A large bird, called " Arruk," is very common: they are usually seen standing on the top of a thick bushy tree, their heads and necks ducking up and down, whilst they utter a sound like the heavy creak of an old-fashioned pump. We are apt to think that the natives of South America (not aboriginal) are utterly ignorant: the men in the little craft could all read and write. I always got Pedro, the mestizo lad, to write me the native names of the fish and trees. One hardly expects to find such a pitch of education in fellows with shirts like the rags remaining to us as relics of the Waterloo standards! Our midday halt was chosen under the shadow of a bank somewhat more elevated than is usual in these low lands. As soon as

my companions had partaken of their light meal,
they severally disposed themselves for a snooze;
not feeling similarly inclined, my eyes wandered
along the bank, enjoying, half unconsciously, the
grace of the motion of the wild canes, that waved
gently in the immense breadth of the tropical sunlight. Soothed at last into drowsiness by the
rippling sea-breeze, the soft continuous whisper of
the canes seemed to lose identity with its cause,
and become human. I heard words fluttering on
the air, and names were murmured belonging to
dear ones far away; but anon the fairy spell was
broken, for the breeze grew fresh, and the swaying
of the canes became boisterous, as the men arose
from their siesta to shake out the old sail again, and
we once more sped on our way.

CHAPTER II.

JANUARY 17TH.—This morning we reached the first of the mongrel Spanish settlements, indicated at a distance by a few young coco trees, bread-fruit, and plantains. At this plantation, which belongs to an old coloured Trinidad man, we were hospitably entertained. Amongst the good cheer was an excellent fish, called barba de tigre. Here plantains are cultivated abundantly. We remained during the day at this place, which is called Cuti-pita, where we found many Spaniards in hiding, to escape from military service in the revolutionary disturbances on the savannahs of Venezuela to the north. The old Trinidad man had numerous Indians of the Guarrauno race in his employ; they were crushing his sugar-cane to make papalon, the moulded sugar of Venezuela, which has a ready sale at Ciudad Bolivar. The old fellow also had some uncoloured rum of very fine flavour, and he seemed to think a great deal of it.

I took a portrait sketch of one of the Indians.

19TH.—We continued our course. There were now sand-spits at each bend. We passed into the

Macarco branch. The bank was well settled for this part of the country.

In the gloaming of a beautiful evening we sailed out into the main river, leaving the channels of the delta behind us. The river is here studded by four islands.

20TH.—As soon as it was light we had a good view of the high lands to the south, the Isnataca range. There are so many bends and low islands, that one is not here impressed, as might be expected, by the really grand proportions of the Orinoco. During the night we coasted the large isle of Tortola, and passed the town of Barrancas. In the dry seasons the breeze blows from the coast. We now went steadily along before it. The hills closely approach the river on the south side, at a point where there are some rocks called Morocoto. From thence to the Sorondo headland, wooded, gently undulating hills rise from the savannah lands, and appear eminently adapted for settlements; for they are a combination of pasture and planting grounds. Cresting some large boulder rocks, and embedded in brushwood, the curious old fortified Spanish settlement, called Old Guayana, rose before us; we noted its cottage-like houses as we sailed under it. It was emblematic of the picturesque decrepitude of the Spanish main. On the other bank the famous Llanos stretch away hundreds of miles towards Caracas, and the great savannahs of the Apure. On the Guayana side are thick woods and occasional savannah. Wild fowl had latterly been abundant. At night we cast anchor at the so-called puerto, Las Tablas. It was a fine moonlight. Next morning I went on shore; the settlement is a wretched place, built on the top of a sloping bank of sand and rock, backed by scrubby brushwoods. It is near the mouth of the Caroni, and owes its existence to the traffic with the newly discovered and rich gold-mines of the Caratel district. Following the right bank for a considerable distance, unmixed with the turbid Orinoco, the clear dark water of the Caroni was to be distinguished. As yet I had seen or heard nothing of "the Southerners" said to be settled in this neighbourhood. Owing to the sandy nature of the soil, these woods are stunted in comparison with the forests of the other parts of tropical America. It is not until you reach the upper lands, which by their rising cause the rapids of the main river and its tributaries, that you meet with the true Guayana forest. In the forests of the upper Orinoco, from the cataracts of Maypures to the mouth of the Cassiquiare, I afterwards met with the most enormous trees I had ever beheld. They were usually of the kinds called by the natives Cachicamo and Jasapas.

For the last two nights I had enjoyed delightful sleep; it was almost like an escape from "Inferno" to get clear of the mosquitos of the delta. We passed sundry rocky islets and boulders as we sailed along; their surfaces were black and shining, from the alternate action of the sun and Orinoco water.

to face page 14.

MOON RISE. ORINOCO

At frequent intervals at this season long flats of sand are exposed. Some parts of the Guayana side have quite an European appearance—one long ridge behind another, the timber apparently low and scattered. On the under side of the limbs of some trees hangs a kind of moss, which floats out on the wind like a mane.

At sunset we passed Guarampo, a place like two rounded hills, and three islands of rock and sand. During the night we ran on a shoal, and experienced some difficulty in getting off again. Heavy rain fell after the moon had gone down; however, wrapped in my waterproof poncho, I cared little for a moderate amount of damping. In January many of the trees are bare of leaves, giving quite a dried-up look to some districts. From the rocky nature of the river beds, the water off some of the points appeared boiling. About noon we passed betwixt the shining black rocks, Rosario and Rosarila; then a dome-shaped mass, called Conejo, which was crowned with cactus. All these rocks are alike, black and shining. I can only compare their lustre to a well-brushed fire-stove, and they give a peculiar character to the scenery of this river. We saw very few " caymans," alligators, of any size. We passed a fine, trim schooner (the first sign of trade yet seen) cautiously bearing down against the east breeze. I noticed that the top of every sandy bank near a hut was planted with " pateas," a kind of water melon.

At sundown, sailing by the point of Angosturita,

we approached Angostura (the straits), or, as it is now called, Ciudad Bolivar, after the revolutionary hero of the country. This is the chief, or I should say, the only town of Spanish Guayana, and is built on a low hill, backed by sandy savannah lands, and flanked by a lagoon. It looked picturesque, with its painted cathedral and flat-topped houses standing out in relief against the clear evening sky.

23RD.—I went to the Government House, through rough-paved, but clean streets, and was very kindly received by the Governor, Dalla Costa. Like every one else here, he seemed to thinkmuch of the new gold-mines of the Ciudad district beyond Upala. He gave me a sort of recommendatory paper, which I afterwards found very useful with the "prefectos" of the interior villages. Fish abound in the Orinoco, some of great delicacy, others larger, but good eating. The smaller kinds would afford capital sport to a fly-rod.

Two fine German or Dutch brigs lay off the town, also an American vessel, and two or three fore-and-aft schooners, in the salt trade of the Araya peninsula, and the cattle exportation with Demerara and Trinidad.

Being obliged to stay here for some time, the hours hung heavily, as the neighbouring country is not particularly interesting. On principle, I took my daily constitutional (I believe exercise is even more essential to health in a tropical than in a cold country) away from the town, over the stony paths,

betwixt hedges of tall cactus and sparse, dried-up
bushes. Some of the low places following water-
courses were green and pretty enough. The
natives call them the "manrichal," from the manriche palms that adorn them. The usual living
things to be encountered, besides the ever-present
Zamora vultures (the "turkey-buzzard" of the
Yankees), are flocks of ground-doves and paro-
quets, and a few gay-coloured finches. There
are four species of tropical starlings common on the
Orinoco, all lively in habits, even to a greater
degree than our starling. Some of them are good
songsters; they live generally in societies. One,
the most striking in appearance, is clothed in
yellow and black, brought into vivid contrast on the
back; he also rejoices in blue eyes and a cream-
coloured bill. The other kinds have jackets of
orange, yellow, and black, in various proportions,
yellow preponderating, with the exception of one
in shining black, enlivened by light eyes of the
greatest brilliancy. The tail feathers take a singular wedge shape, something like a common barn-
door fowl, but the point of the wedge is directed
downwards instead of up.

The number of ants visible everywhere is one of
the features of this part of the world. As I was
sitting under the big Cieba, or milk cotton-trees,
reading, as I usually did in the mornings, I saw
a great instance of the sagacity of these little
creatures. The entrance to their excavation was
on the top of the river bank, in a place exposed to

the heavy squalls; therefore, when working below, they piled the grains of earth, brought to the surface in their mouths, well to leeward, to prevent their being blown down into the hole again. These ants were of the small species, and give out an aromatic smell if crushed. Some sultry afternoons later in the season, I saw numbers of gigantic individuals of the leaf-carrying Saüba species, "biscachos." They were winged females, but their beautiful transparent wings are only the gift of an hour; for after the first use of them, which enables the insect to rise on the breeze, and be wafted to a distance, they either drop off or are detached. Thus she becomes the progenitor of a new colony of her kind. The greatest flights occur during the heavy evenings at the beginning of the rainy season. These ants, when full grown, are an inch in length, and when on the wing, they resemble the wasps called "Jack Spaniards" in the West Indies. Once alighted after their journey, and finding it difficult, if not impossible, to rise again, they deliberately stood on their heads, and brushed off their wings with their hind legs. Their bodies were full of eggs. Often as they came heavily to the ground they were seized and devoured by a swift-footed lizard. This lizard has a peculiar trick of nodding its head, in the manner of the toy jack-asses of the Lowther Arcade. They dart from the crannies of the buttress-like roots of the Cieba trees. All the hens in the neighbourhood seemed also on the look out for dainties, and ran after the flights of insects as they sank earthward, with outstretched necks.

The only hotel is, as usual in Venezuela, kept by Germans. I swung my hammock in a large stone-flagged room; the charge, inclusive of meals, was two dollars a day. The scenery, after one attains a certain elevation, although not altogether beautiful, is singularly majestic; it seems so evident that a great continent, with rivers, forests, sandy deserts, plains, all on the grandest scale, is opening to view. The soil is almost all sand; and what is called by Humboldt, Amphibolic rock, I believe, everywhere forms the surface, except in favoured hollows, where grow plenty of cachew and mangoe. The bird of sweetest note here is robed in sober grey; his song' somewhat reminded me of our thrush. In this hemisphere, it is not so much the height of the temperature that is felt, but its continued heat all the year round.

FEBRUARY 11TH.—I procured a boat from Señor Antonio Dalla Costa for my purposed trip to the Caura. She had once been the life-boat of a steamer, but having been long exposed to the sun and rain, Rogers and I were at work on her for some time: being of iron, I found that our skill would not suffice to put her in order; and, after all, I was forced to abandon her. As I was thus compelled to prolong my stay, I removed my hammock to the house of an American woman, one of the last of the southern settlers, who came two years before. Nothing so impresses upon a traveller the great distance, in these latitudes, separating him from home, as the starry face of the sky—the polar star, planted so high at home, here just topping the trees. The sun rises and sets somewhere about six o'clock. One scarcely dare venture a swim in the river at Angostura. I did so at first, till assured of the probable great danger from the "tembladors" (electric eels). A shock from an eel would send a bather to the bottom without reprieve: this was a tremendous denial in such a climate; however, there was the alternative of bathing in the shallow sandy creek beyond the town, with a "tutuma" or calabash, in the cool of the mornings. Poor Rogers was struck by a "raya," "sting-ray," whilst wading in the shallow water at the brink of the river, and suffered considerably; his leg swelled, and he was rendered almost incapable of walking for some twenty-four hours. A few Indians of the painted tribes from the interior are occasionally to be seen lounging about the stores; but the aborigines usually frequenting the town are Caribs, still robed in the deep blue cloth described by Humboldt, and are the most important of the native tribes: in person, they are fine men, of full coppery complexion.

They inhabit the lands lying between this river and the old missions of the north coast and Piritu : Governor Dalla Costa estimated their number at not less than 40,000. Angostura relies largely upon the Caribs for its supply of cassave, the bread-stuff of the population: this they bring from their conucos for sale in great flat cakes. A branch of the race living about the head of the Pomeroon and Caroni dress in paint and feathers, much as the other Guayana Indians. The inhabitants of Ciudad Bolivar are a conglomeration of Spanish, Indian, and Negro; but the merchants of the town are chiefly Germans. The roads and Government works are made and kept in order by a chain-gang of the criminal prisoners, there being no heavier punishment for crime. The convicts are brought out every morning to work under guard, and a more villanous-looking collection of different types of men I think I never beheld. Among them was a low-class Frenchman, who, associated with a repulsive visaged Negro, had long been in the habit of robbing and murdering travellers on the road from the Caratel mines. On the quay, one day, some soldiers ran past, loading their old flint-lock muskets as they went, halting occasionally to level a shot at a man who had just left a pulperia, and was making up the road. The object of pursuit (but that day released from a term of imprisonment) had found his former sweet-heart at the pulperia in the company of a rival, and overpowered by jealousy, had run her through with a machete. The miscreant was a tall Negro, who had been notorious as a bully among his fellow convicts; he was ultimately severely wounded and captured.

I did not long taste the hospitality of my American hostess; she was not blessed with a particularly amiable temper, and she kept a stock of some dozen parrots in readiness for a Yankee skipper, who traded with New York. Imagine the having to dine and carry on a correspondence in the parrot-house of the Zoological Gardens, and then an idea can be formed of the continual riot I was expected to endure patiently.

I shifted my quarters to the house of a good old Barbadoes woman, who was quite motherly to Rogers when he had the fever. The domicile of Mother Saidy was a sort of reunion for all the niggers from the British West India Islands, where they met to discuss affairs private and political. It was most amusing to see what pride they took in being British subjects, and the contempt in which they held their dark brothers of the main. Mother Saidy had also a weakness for picking up, and caring for stray chicks of doubtful pedigree.

The affairs of the Venezuela Company appeared to be in a deplorable condition: the several batches of American emigrants (southerners), when they arrived, found no preparation had been made for their reception, nor was there a competent agency to direct their movements; the consequence was, that discovering much of what they had been led to expect untrue, they became disgusted with the country, and, instead of keeping together for mutual assistance, many of the younger men went to the mines, and the others dispersed in different directions. From what I heard of them, they seem to have been anything but well-selected, respectable men, and, for the most part, the offspring and refuse of the long civil war. The sites chosen on the Paraguay (tributary of the Caroni), for those who did not afterwards come provided with implements and materials for farming, were singularly unhappy, as the sole means of communication with Ciudad Bolivar is by donkeys, and that sorry mode of conveyance difficult to obtain. Had the unfortunate people been located up the main river in the vicinity of the Caura, and had they combined interests, they might have effected something. Leaving out the consideration of the expense, no reliance can be placed on the native labour: a colony must, to prosper, be self-supporting and independent; besides, Venezuelan morals are so despicably corrupt, that I think no Englishman or American would like to introduce the women of his household within hearing of the common every-day parlance. If the settlers had gone up the stream, they would have gained the advantage of easy access to the market at Bolivar; even if they were destitute of boats, their produce could have been rafted down to the town. Should the Venezuela Company not be able to organize their next emigration better, it would be wiser to abandon the design in the beginning, as a failure must invariably end in the disappointment and distress of all engaged therein. It is much to be desired that the published statements as to the easiness of procuring labourers for the farms, &c., should be corrected, as they are calculated to raise false impressions and expectations: at the same time, there can be no doubt of the natural richness of the soil in tropical produce.

19TH.—With the assistance of Mr. Derbyshire, an English trader on the river, I engaged an experienced pilot, named Ventura, and purchased a fast little native-built lancha, as I purposed exploring the Caura in search of india-rubber. Angostura is situate in that drier district of South America which extends from the Cumana coast, embracing the Llanos and part of the lower Orinoco. The prevailing appearance of drought is in the dry season heightened by the surrounding savannahs, the surfaces of which are broken only by rocky ridges, with stunted cachew and chapparo trees, and bushy thickets, often burnt by fire.

On Monday, February 22nd, we sailed from Ciudad Bolivar early, running as far as Almacen, a little hamlet on the south bank.

23RD.—The Orinoco is here near its lowest, and it was difficult to keep the channels between the sand-banks; having run aground, we got off with some trouble, and I made the lancha fast for the night in a small basin betwixt the rocks, near the village, or pueblo, of Borboa. I despatched Rogers and Ventura there, as we stood in need of a canoe curiara; they were unable to obtain anything of the sort, and I remained for the night on the sand.

24TH.—The lancha sailed fast before the east wind; the breeze was strong; so fearing shoals, we reefed the main-sail. Passed the mouth of the Rio

Aro; at this season only navigable for canoes. The timber on the immediate banks is small, but many of the tree-tops, rising above the general line of the woods, are very graceful in form. At mid-day we reached the Isla and Roca del Pao, where I put in for the night, at the village Muitaca, in order to try again for a canoe. The view in ascending the river was very fine.

25TH.—Having succeeded in obtaining a curiara, we set sail about noon, came to anchor near the great bend of the river called the Torno, where the Orinoco is turned from its last course by a knot of hills on the Guayana side, and round which it makes a sharp detour to windward, troublesome in sailing up stream.

For three days we were prevented from rounding the Torno, and were obliged to work a part of the fourth, although it was Sunday.

I lost my anchor, which was ultimately the cause of much vexation. A bend seen from the river reminded me of the black hills of South Wales; it was the end of the line of hills, and occasioned us our double labour.

MARCH 1ST.—Passed Roca del Inferno, a place where the river is compressed betwixt high rocky shores, and divided by an island. We stopped at the pueblo of Mapire, situated on the left bank, to lay in a stock of the cassave bread.

2ND.—We spent the night on a sand-spit, not far from the mouth of the Caura, and I shot the largest Muscovy drake I ever saw here : being only wounded, he led me a long journey through a shallow, muddy lagoon, before securing him.

Four kinds of ducks are to be had in great abundance, especially during the dry season—the large green Muscovy, the long-legged "caratero," a smaller red-legged species, and a kind of teal, "quiriri." The long-necked darter or "cortua" is plentiful, and much prized for its fleshy substantial proportions—an important consideration when the appetites of a large party are concerned. When shooting in the mangrove swamp on the sea-coasts, I had found this bird so intolerably rank as to be quite uneatable; but here on the Orinoco it proved excellent fare, and resembled goose in flavour.

We got within the mouth of the Caura in the morning, but were frequently aground from the lowness of the river. After proceeding about a mile, the men said we could go no further in the lancha, in which dilemma I insisted upon making every trial possible, and at length we discovered a tortuous channel, with sufficient depth of water, winding in and out the sand-banks.

4TH.—The wind favouring us, we pushed on again.

On the Caura, the sand-flies were exceedingly annoying during the day, but at night we slept free from mosquito attacks. The banks soon became high, and are here composed of white clay. A pair of cranes had built their nest (which looked like a basket of sticks lodged in the central fork of

the branches) in a great, spreading Cieba tree.
The flesh of the green Soldado crane of the Orinoco
is excellent. The appellation of " soldiers" is very
appropriate, when these cranes are descried through
the mists of the dawn on the river, or drawn up on
moonlight nights on the playas, or sand-banks, as if
in motionless ranks, like those of an army in readiness for the encounter with the enemy. About
noon we had a glimpse of hills, then passed a little
creek from the last leading to the pueblo (Indian
village) of San Pedro. We sailed on slowly till
afternoon, then camped at an island called Los
Tembladors (electric eels).

5TH to 16TH.—We had much trouble with the
lancha, standing up to the middle in water, vainly
trying to coax her into some channel. I determined to push on in the curiara with Ventura, who
could paddle well, leaving Rogers in charge behind.
Paddling until afternoon, we reached an inhabited
place : we moored the curiara here, and struck out
for the country, in order to avoid the windings
of the river : crossing first the belt of wood
skirting the water, and then marching over a high
savannah country, in some places recently burnt.
We waded through a small stream, the water of
which looked dark in the starlight, with its dusky
girdle of tall palm trees. We slept an hour or two
at an empty house we discovered on the other side
of the stream, and again pushed on across the
burnt brushwood, reaching the pueblo of Maripa
soon after dawn, tired with our march. As the sun

rose, the view of the blue hills over the river
was beautiful; on the other hand, a great savannah
stretched away indistinctly in the mists of the
morning like a sea, the line of horizon broken only
in one direction by the summits of a cluster of
mountains, probably the same which cause the
first rapids of the Caura; the level of the plain
rarely, and only broken by an occasional range
of Manriche palms, with little thickets about their
bases, in indication of some watercourse.

The prefecto of Maripa was absent from the
place, collecting sarapia (*Dipteryx odorata*) (tonka
bean), which appears to be the chief thing looked
after on the Caura by the lazy natives. They are
able to collect enough during the season to enable
them to obtain all the necessaries they require for
the year, not even cultivating cassave, or frijoles,
but relying for provisions on their neighbours up
and down stream. The woods abound in medicinal
trees. A variety of quinine is said to be found
here. Generally the banks of the lower course
of the Caura are fringed by a belt of forest, more or
less· broad; then the ground rises considerably,
and stretches away in the sandy savannahs. Sometimes, however, the savannah ridge descends
abruptly to the river, forming a bluff. As a rule,
the forest belt gradually broadens as the raudales
are approached, and is wider at Aripão than at
Maripa.

I walked to Aripão, the next pueblo up the river,
with Ventura; like Maripa, it is situated on the

high sandy savannah away back from the river, and the plague of sand-flies. It has some fifty men negroes, descendants of runaway slaves from the old British settlements on the Demerara. With the assistance of the Spanish prefecto, I hired a large canoe, to bring up the demi-johns I had in the lancha for collecting balsam.

From Aripão there is a beautiful view of the Hilaria Hills on the other side of the river. Not being able to procure the necessary provisions, I was compelled to return immediately to Maripa. Ventura and I had only eaten frijoles (beans), and hoping for a reinforcement for our larder, were on our way down stream before day.

8TH.—In the evening we reached the lancha, and found all well with Rogers. Having secured the lancha fore and aft in a deep pool near Los Tembladors, we conveyed everything from her to Maripa, my purpose being to leave her here until I could bring something to freight her,—balsam capivi, &c.,—and then return to Angostura for the season.

11TH.—We reached Maripa early, and saw the prefecto, a very tall black man, of singularly prepossessing appearance. On the strength of my official letter, he bestowed sufficient provisions upon me for our ascent of the river, and promised me an introduction to an Indian " Capitan." I purchased a large bull for about £4, no meat-provender being forthcoming otherwise: this delayed me for two days to dry the meat, after salting it.

The river, after rising several feet, went down as much. I obtained two Indians from San Pedro, the village of the Arigua, a remnant of a once numerous tribe. The Capitan of the pueblo accompanied me up the river, to recommend me to a chief who was a relative of his.

15TH.—Ventura was very drunk, having induced me to give him an advance. As he interfered with the start, I had him stowed away at the bottom of the canoe as ballast. We only got as far as the puerto of Aripão, where we camped. Here I bought a good curiara for thirty pesos. I had lightened my baggage as much as possible, in order to make the most speed up the rapids.

16TH.—Camped on the left bank: many people were collecting sarapiá.

17TH.—Started about ten; came to the mouth of the Mato. The woods continued free from much undergrowth, and were now interspersed with cacurito, zagua (wine), and other palms.

The water of this river is coffee-coloured in shadow, and reflects objects on the banks very distinctly.

18TH.—We had a first experience of a wet camp on the river, for it rained all night. My mode of travelling was to take tea and "a bite" (literally) of cassave, or something, and be under weigh by sunrise. Halt for one or two hours, and breakfast at mid-day. Camp about an hour before sundown, or earlier, if a convenient spot offered. We halted to-day near the end of Isla Larga. Great-domed rocks now rose from the bed of the river. We entered the first rapids of the Caura, called Mura, which at this season are very dry, and the flow, therefore, is not so great. The pools among the stones of these raudales were full of an elegantly shaped shell. A species of cicada makes a tuneful sound in the neighbouring woods, which closely resembles the jingling of little bells; another approximates in sound to the whistle of a lively locomotive. Whilst the camping preparations were in progress, great masses of cloud rolled up from behind the forest, well harmonizing in grandeur with the mighty boulder-stones of the raudales. We were deluged with rain for the greater part of the night. The best course to pursue in camp during such weather is to stretch a cord above your hammock, after it is slung between two trees; over this cast a rug, which will give you a tent-roof, and you may sleep underneath it in defiance of any amount of rain.

CHAPTER III.

MARCH 19TH.—In the morning we proceeded up amid the great water-worn boulders, which form the Mura raudales. After our halt, the river became somewhat cleared of obstructions. I tasted the fruit which contains the sarapia or tonka-bean for the first time ; moistened with water it is very refreshing. The graceful Manac palm is also to be seen on these banks, and the zagua. Camped for the night at the edge of the beautiful woods which clothe the hills at the foot of the Piritu rapids.

20TH.—A wet camp. We passed up the raudales of Piritu; the woods looked very inviting, the foliage of the trees beautifully varied. There were many sarapia and balsam capivi; those nearest the bank were festooned with the graceful levaina creeper. At noon, we arrived at the first purely Indian settlement on the Caura, Chapparo, the head man and men of the village were absent up the river, so that I was compelled to wait for their return, there being no other alternative.

21ST.—Palm Sunday. Chapparo was so infested with niguas, that the Indians were obliged to sleep elsewhere. When evening came, we, together with the remaining inhabitants, embarked in the curiaras,

to face page 32.

ORIGUA VILLAGE, UPPER CAURA.

and paddling over to one of the little rocky islets, swung our hammocks in the clumps of trees occupying the centre. There the nights were pleasant enough; as it was the dry season, it seldom rained, and owing to the huge masses of sun-heated rock, the air was of a very uniform temperature.

Here, at Chapparo, I found myself in a purely aboriginal society, with the same primitive manners and almost the same personal habits as among several of the remoter tribes in Central America. These Indians had fine clearings, planted with cassave (manioc), sugar-cane, &c.; but they do not care to cultivate fruit, with the exception of a few pine-apples and bananas. One of the little Indian girls had flaxen hair, but bore no other European resemblance.

25th.—I occupied the time in searching the forest in all directions for sarapia; trees abounded, but they bore little or no fruit in this district. The balsam trees are easily recognized by their peculiar light green foliage, and the light colour of the smooth bark of their limbs; the sarapia, on the contrary, has heavy, dark green leaves; it is said to bear abundantly only on alternate years: the fruit of the sarapia, of the Caura and Cuchivero is largely consumed by the Indians; the stone contains the tonka-bean of commerce. The variety of the sarapia I afterwards saw on the Rio Negro of Brazil, at the rapids of San Gabriel, and elsewhere, was not edible. In my tramps through the forest, I always carry my double-barrel gun, and I used to

D

make Ventura sling my short Snider across his
shoulders. I am sure, had he come under the
notice of any civilized person, he would have been
considered a most desperate character, although he
was really anything but formidable. A great deal
of his fierceness lay in his unkempt, lank, black
hair and beard, and a barber would soon have
denuded him of his apparent ferocity.

The Caura above the settlement is much divided
by islands, and the channels between them are
impeded by rocky runs. In the wet season this
portion of the river must be impassable. The left
banks' are held by the Taparitos, a race of Indios
bravos, hostile Indians, of whom the scattered
Arigua stand in considerable awe. There does not
appear to be much actual fighting between them,
however, as an alarm of their approach occasions
a general bolt into the bush. The Taparitos then
content themselves with planting a few arrows in
the deserted houses as a defiance.

25TH.—GOOD FRIDAY.—The night before, an
alarm of the hostile Taparitos excited our island
camp, by the appearance of a strange fire on the
bank opposite, which was an uninhabited forest;
but as it soon vanished, it must have been a will-
o'-the-wisp. The old Capitan was away fishing, as
usual, having left the women folk in our charge.
When in the wilderness, my gun is always ready
for decisive action, and I told Roger to get out
a few Boxer cartridges for the Snider.

Although we had no rain, this morning the

heavens were draped with heavy cloud curtains.

27th.—EASTER SUNDAY.—The tree called Floramarillo was foliaged (no leaves) with flowers of the brightest yellow; the wood of this tree is very hard, yellow, with a dark heart,—from the latter the Indians make bows: the flower remains but a day or two, and then gives place to green leaves. The lofty tree called Cooti, now lost its heavy bunches of violet blossoms, which before had rendered it so conspicuous among the other forest trees. The Cieba, or milk cotton-tree, grows to an enormous height and girth, throwing out great flanking buttresses, between which there are often black-looking cavities, appropriate lodgings for the great Tigre of these solitudes. One day, whilst resting on the rocks by the water-side, after a hard morning's work, cutting black-wood in the forest, we saw a curiara full of men making for the settlement: they proved to be a party of Guangomo, from the source of the Caura. They were armed with the *blow-tube* as well as the bow, and were adorned with large coronets of parrot and macaw feathers. The Arigua Indians understood nothing of their tongue; they awaited the head-man, whom they had come to see in return for his visit to their settlement. I certainly never saw Indians paddle more vigorously, though their arms were tightly bandaged just below the shoulder, causing the muscles of the upper arm to swell out considerably; but it did not appear to inconvenience or retard them in the least. They wore ornaments of wood sticking through the under lip.

APRIL 1ST.—The men of the settlement not having come back, and as I could not get sufficient people to explore the river for india-rubber, I determined to go higher up to the next serra, where there was a probability of plenty of game and fish. The old Capitan of the Arigua and his son accompanied me. After starting, we first came to the little river Cani, where we found the party of Guangomo engaged in making a curiara from a large sasafras tree. They were burly fellows, with deep copper-coloured skins. In feature they were anything but handsome. I have never seen as strikingly handsome faces in South America as some of the Central American indigenes. These Indians seemed to have a great partiality for tame animals. This was only a travelling party, yet they had parrots and macaws about the camp. They lent me their long, light canoe in exchange for mine, to be returned when next we met, as mine was heavy, and little suited to the present journey. Next day, we passed Serra Cangrejo early: it is situated on the right bank, and clothed with a magnificent forest, and the vapours of a rainy night were rolling away up its side. It is wonderful what majesty is sometimes lent to a scene in the tropics by the rolling cloud and the mist. The woods above the Cani have a truly gorgeous appearance, owing to the number of picti and zagua palms. In the afternoon of the preceding day we had a taste of the rains; but although clouds of rain and vapour passed over us, we had a roaring fire in camp.

Leaving the main river, we entered the Nicare. Our first evening camp on this tributary was pitched at a short distance from the mouth. We had a monstrous supper—panji (a sort of wild turkey), heron, ibis, electric eel, and sting-ray.

2ND.—We continued our progress up the stream, and came to an afternoon halt; El Capitan going fishing, whilst we made preparation for the night. The water of the Nicare is more like that of the Orinoco than the clear brown tide of the Caura. This branch river seems a favourite locality with the macaw. I saw several pairs with nests in the trees on the banks. They make their nests in a hole high up in a hollow tree. Through the opening of one of these holes I caught the bright plumage of the hen, while the male bird occupied an adjacent bough. The forest on the Nicare is, for the most part, remarkably clear, and wild plantains are very numerous, and grow to a great height.

The stream is famous for excellent fish. In the morning I had been into the forest with my gun; after shooting a few birds, a feeling of giddy faintness came over me, accompanied by a disagreeable sensation of doubt as to whether I should be able to get back to the camp; however, shaking myself together, I made an effort, and succeeded in reaching that destination just before the fever obtained mastery over my limbs. I remember at the time connecting this sudden seizure with the effluvia of a tree that I had notched for the purpose of examining the wood.

5TH.—I broke up our fishing-camp, and started down to the Arigua village. These Indians have a pale skin. Their language, like that of some of the other American tribes, has the singular appearance of the words being enunciated with difficulty, as if a slight stutter incommoded the freedom of the tongue; that of the Guangomo, on the contrary, is loud and harsh.

15TH.—We were obliged to start down stream, as we were all (Ventura and Rogers as well as myself) touched with fever.

24TH.—Halted for several days at Maripa, to recruit a little strength. In the forced journey from Arigua to this place, we suffered dreadfully from exposure to the sun whilst the fever was upon us. Ventura alone was able to use a paddle, so after passing the rapids, we made the three curiaras fast together, and dropped down at mid-stream. We were compelled to continue throughout the day, as we were quite out of provisions. The only manner in which I could obtain anything approaching to relief, was to keep a towel constantly saturated with water over my head. Also, after the paroxysm of fever had abated, I would, during a halt in the cool of the morning or evening, drag myself to the brink of the river, and lay myself down in the rippling water. At Maripa I got two Indians

to face page 38.

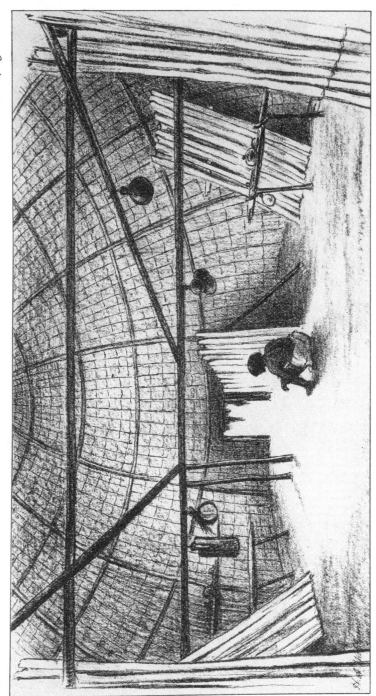

Interior of an Arigua Lodge, Upper Caura River.

from the village of San Pedro. When we reached
Los Tembladors, where I had left my lancha, I
found that during my absence on the upper river,
some of the rascally Venezuelans who frequent the
neighbourhood of the Mato Hills during the sarapia
season, had stolen the hawser by which she had
been made fast. We discovered her stranded on
her side, left by the receding river; the upper
portion had become so warped, and the planks
so opened by the sun of the dry season, that we
were obliged to abandon her, and continue on in
the three canoes. We halted only once more on
the Caura, in order to visit a reedy lagoon in search
of game. We wounded one deer, which got off, but
we secured plenty of ducks: the water was nearly
dried up, with the exception of a pool or two near
the centre of the lagoon. These were literally full
of small turtles. Ventura quickly denuded himself
of his nether garments, and tying a piece of string
round the end of each leg, he soon filled them with
little turtles; he presented a ridiculous spectacle,
capering about through the deep black mud in
chase of the scattering reptiles and wounded ducks.
When he marched back to the canoe, he looked
exactly as if he had on a pair of dust-coloured
tights.

On the Orinoco during the dry season heavy
squalls are frequent, and the only warning given is
a sudden lowering of the sky in the direction of the
danger. A few days after we had left the Caura,
we were very nearly coming to grief in one of

these temporals. We were coasting a long line of abrupt bank or bluff, called barrancas, when the experienced eyes of my two Arigua caught a warning from the clouds; it then, of course, became a race for it, for if a canoe were caught in a temporal against a barranca, it would inevitably be swamped. We just managed to gain a place where shelving sand stretched down from the bluff, and to run the head of the canoes on the sands; but ere the Indians could draw them fairly up, the water curling before the driving squall washed clean over them. They filled directly. However, we held on, and so prevented their being washed away, and in half an hour all was over. I had my two cases of Eley's cartridges lifted out and brought on shore to be dried in the sun, little expecting that they would ever be fit for anything again, as they had been a considerable time under water in the canoe. When they were dry, I was agreeably surprised to find that they went off as if nothing particular had occurred to them. We got safely back to Angostura on the morning of the 8th of May, having left Maripa on the 25th of April.

The month of May was excessively dry and hot at Angostura; but the Orinoco (the upper course of which travels amid humid forests, where the air is ever laden with vapours) continued to rise steadily, bearing along on its bosom uprooted trees. The stony hills, perhaps I should rather say ridges, opposite the town had, within a few showery days, become so modified in their outline by the fresh verdure of the small trees, as to be, if not exactly picturesque, at any rate pleasing to look upon.

The small trees of these ridges of the lower Orinoco are generally bare of leaf during the dry season.   A very curious kind of cactus is here to be seen in societies; it is cylindrical and domed in form, but does not attain more than a foot and a half in height.   Its fluted edges are thickly studded with a cluster of sharp spines: on its top it has a boss of silky substance, somewhat resembling cotton wool; from this it produces delicate little crimson flowers, which pass away with the freshness of the morning, in a day or so to give place to a small fruit, in taste like a hedge-row strawberry.   It is curious to see the flowers and fruit come up out of what appears to be only dry, silky cotton.

One of the most noticeable natural features of this part of the country is the changing course of the wind (the Trades).

Blow it strongly, breathe it softly—

it is still from the same north-easterly quarter.   A calm is always succeeded at this season by a squall of wind and rain.

On the 4th of June the first heavy fall of rain caused vegetation to spring up so rapidly on the sandy ground, so dry before, that we had difficulty in recognizing places we were familiar with.   Delicately blossomed plants sprang up wherever the ground was shaded by cachew, mimosa, and mangoe trees, in the neighbourhood of the rapid little sandy stream running between the stony hill on which the Ciudad is situated, and the bluffs of the great savannah beyond. Very many of the flowers were violet of various shades.

Ciudad Bolivar has a fine market near the river; but though the market-place is good, it is miserably appointed: the most prominent objects are two enormously fat women, who have stalls, one on each side of the entrance. I think these individuals must nearly approach the proportions of those royal beauties described by travellers in Central Africa.

I again took my meals at the house of my old friend, the good-natured Barbadoes woman, Mother Saidy. After I had been staying with her, she evinced such a partiality for me, that I found it difficult to discover when she required money for marketing, unless Rogers undertook the office of finding out the necessity existing.

Owing to the loss of my lancha, and the consequent collapse of my projects on the Caura, I determined to make a push for the Amazon valley, by way of the raudales of the Orinoco. Watkins, a young Southerner, desired to join me. We arranged that we should pass the Orinoco Cataracts of Atures and Maypures, and thus gain the Rio Negro, and thence the Amazon, either by the Cassiquiare or the Atabapo. Watkins and I were in good condition, and did not fear as to the result. My new comrade had seen much rough service in the late American war and in New Mexico, and had just arrived at Angostura, walking all the way across the plains from Valentia and Caracos. Señor Dalla Costa kindly let me have my old Caura curiara back again.

CHAPTER IV.

AUGUST 6TH.—I took the steering paddle, and started, with Rogers, Watkins, and an Indian, for the Rio Negro. Watkins and Rogers were in great spirits, and I, of course, did not dwell much upon the impending difficulties of the way. Señor Dalla Costa gave me letters to the governor of Amazonas, as the Venezuelans call their south-west frontier district. As it was the starting day, we paddled only to about the distance of ten miles above the town, and there camped. The rain was nearly at its highest, and in consequence the current was very strong.

7TH.—This morning we had been travelling quite among the tree-tops and wild calabash bushes, as the lowlands were under water. We camped by three o'clock; found plenty of iguana in the trees. I now began to see that I was fortunate in my Indian (Ramon); he worked willingly and well, and, in his Indian way, did things quietly, without my having to remind him of them.

8TH.—Sunday in camp. The woods had much improved since the rains; everything therein was now green and graceful. On the side of the stony hill behind our camp were many of the trailing cacti, bearing plenty of their pleasant fruit. Mosquitos swarmed annoyingly, and I feared would increase in numbers until we had passed the raudales.

9TH.—Thanks to my mosquito curtains, I had a good night; but my rug was lost, unfortunately being carried away by some thick bushes which drooped into the curiara. We were all much refreshed by the rest of yesterday at Iguana Camp, as Watkins called it. To-day we camped above the village of Almacen.

10TH.—I was rather put out by the defection of Watkins, who, declaring himself unwell, took leave of the expedition, and returned to Bolivar. As I had rather relied upon him, I was much disappointed. It had been very fine throughout the day, but now (sun-down in camp) there was a threatening squall on the horizon. We camped below Borbon. I fancy I should create a crowd in Regent Street if I were to appear in my canoe costume—red flannel shirt and pijamas, and on my feet a pair of alpragatas, a native cross between sandals and slippers: here, however, the attire was very suitable to all emergencies. Steering even a small curiara such as mine was no light labour, day after day, against stream. I mostly followed the south bank, and dodged in and out among the bushes to avoid the current; therefore the distance covered when we reached the Rio Negro was immense. After dark I was driven from my hammock by a yellow and black beetle; my mosquito curtains were no pro-

tection from the advances of these creatures, for
they nimbly ran up the inside : they made a
curious squeaking noise when disturbed in dis-
cussing the remains of the supper. Their bite is
like the fall of a spark of fire on the skin. I often
encountered this beetle afterwards in the Orinoco,
but never in such irrepressible numbers. We were
obliged to take refuge in the canoe, where I passed
rather an uncomfortable night on the benches.

11TH.—On reaching Borbon I tried to get an
Indian to take Watkins's deserted paddle, but Ramon
could find no substitute in the place. I bought
some tessajo (jerked meat), and then made fast for
the night under the lee of a large floating log, on
which we made tea. At this, the height of the
rainy season, little or no dry land is to be met with
often for several days' journey.

12TH.—A very wet morning: it cleared up at
noon, and we crossed the mouth of the Aro. The
woods here had quite the aspect of our autumnal
English forest colouring, which likeness was
heightened by an over-hanging leaden sky, from
whence descended floods of rain. I missed my lost
rug very much.

We paddled late; the woods were everywhere
flooded, so we made fast to a submerged lavorel
tree, whereon I slung my hammock between two of
the limbs, and with the sweet-smelling wood cut
from its dead branches we made a fire in the fork.

13TH.—During the morning's paddle we came
to a fine conuco, where all hands were engaged in

preparing the cassave (manioc) bread. I bought some; it is very palatable when quite new and hot.

My steering and paddling exertions had rather tired me this morning; for as Rogers was ailing again, the heavy work rested with Ramon and myself; but my Indian was an excellent fellow. I stayed for the night at Muitaco, and a host of the mongrel natives came down to the water when we brought up. Many of them offered for hire; but when they heard that I was bound above the rau-dales, they became less eager; and when I stated that I wished for an Indian, they appeared astonished at my non-appreciation of their creole services; however, I could get no Indian from among them.

14TH.—We skirted the hills of the Torno, by way of a large lagoon of inky water, and lost a whole day by doing so. Sometimes we could with difficulty find our way through a maze of bushy tree-tops; at others, we seemed as if gliding across fields of young corn (a kind of grass shooting up from a considerable depth), over which the pretty little Jacoma lightly ran on their great spreading toes; or a thick belt of the curious Ranwana plants barred the way. Packs of Perros de Agua (a large species of otter), sporting among the submerged trees, gave forth their peculiar mewing cry at being disturbed by the stroke of our paddles, or suddenly lifted their heads and shoulders out of water, in order to reconnoitre us, at the same time displaying a goodly set of sharp white fangs. I shall not take space to relate our difficulties against the swift

current among the trunks of the flooded forest,
through which we eventually gained the main
stream again, and camped on firm ground at the
Torno.

15TH.—I passed the Sunday morning in camp,
as usual. This morning the scene was quite lively.
Ramon and Rogers were busily engaged in cutting
up a fine binado (buck); I had given the Indian my
favourite Holland to look up a few ducks for dinner:
he returned presently to ask for help to bring in the
buck, which he had killed with a charge of number
3 shot. He was much surprised, and highly delighted with the hard, killing qualities of my
escopete. The Sunday's rest was a great relief.

16TH.—Heavy rains. Slept in the rocks opposite Isla Inferno.

17TH.—No dry land. Passed the night uncomfortably in the curiara.

18TH.—Passed the night on some large rounded
slabs of rock at a place called Peña Negra; much
rain.

19TH.—Made a good distance this day's paddle,
and spent the night more comfortably in my
hammock, slung on the limbs of a tigre tree above
the water, at the mouth of a creek, or caño, called
Torno.

20TH. — We reached the neighbourhood of
Purney. I determined to make another attempt to
get an Indian here. I had a pleasant walk across
the savannah, with Ramon as guide, to the pueblo:
the clumps of Chapparo give quite a park-like

appearance to the country. Ridges of dark, rough rock stood out everywhere, and each dip of ground was veined by a little rill of clear water with a marshy border; and springing up out of the coarse clumpy grass, I noticed pretty, daintily delicate little flowers.

I saw the prefecto, a miserably dirty-looking blear-eyed creole, and could obtain from him nothing but promises for "mañana" (to-morrow). No Indians live in the village, their camps being away somewhere on the savannah beyond.

21ST.—As no Indian was forthcoming, I paddled away, having first laid in a good stock of mangoes. The fine clumps of these beautiful fruit trees now seen growing on the savannahs of Purney, no doubt indicate the sites of old Spanish settlements. The soil of these savannahs seems especially suited to the mangoe, as even the produce of the wild trees is of excellent quality.

The mouth of the Caura was calm as a mill-pond, with no current; however, we had hard work to round the first point above.

22ND.—SUNDAY.—Paddled a short distance, in order to find dry footing enough to cook a wild turkey for dinner: the day of rest was most acceptable. I read over old letters. At night we made fast to a bush some way from the bank, to swing clear of the mosquitos (zancudos).

23RD.—There are many good fruits in the bush at this season, with which to flavour guarapo, the native drink of sugar and water. One tastes like

E

a French prune; also the hobo plum. The "plop-plopping" of the latter is to be heard on the quiet waters of nearly every mid-day halt. I slept martially on my gun-case, placed across the thwarts of the curiara. At sun-rise we had the heaviest storm I ever experienced since I camped on Slapping creek, Mosquito coast.

24TH.— Wet dawn again; everything dis-tressingly soaked. We passed some very difficult places, with strong currents, and thorny chinchora bushes. Passed Isla Mictica. The day was very hot, and the river looked like molten silver. For-tunately for my strength, I am able to stand heat very well; but poor Rogers felt it extremely, and, as usual, gave out at half-day. The floating trunk of an immense catchicamo tree formed our chamber for the night.

25TH.—A long paddle; I was very tired. Passed Boccadua Isla. One of the most conspicuous trees on this part of the Orinoco is that called by the natives *Matapalo*. Originally, it is a magnificent parasite; but when once it has embraced the trunk of a forest tree, it mounts higher and higher, till its glowing foliage mingles with, and then tops that of its supporter : its supple limbs, now tightly com-pressed, flatten out, and gradually spread over the whole trunk of its victim, so enclosing it as to deprive it of life, and then it stands self-supported, a great tree, bearing aloft a dark green dome, which casts a grateful shade for a halt in the noon-tide heat. It seems to be a vegetable Anagonda. We cooked on a savannah above a little Carib village, with an unwritable name. The sand-flies pestered us again dreadfully, and I found it almost impossible to scribble a line in peace. After dining, we pushed off and tied to a bush for the night.

26TH.—Made a halt to sun the things on the savannah; had breakfast, and started afresh. We had for the last few days a fair breeze; to-day we put up mast and sail. As we went smoothly along, a curious dull sound came from the water. Ramon said it was caused by a fish. About the middle of the afternoon we rounded the end of Isla Tukuagua, a large island. Here the breeze deserted us, and we resumed the paddles. We had stiff work with them in rounding Punta Bravo, on the upper side of which we were glad to make fast. No mosquitos, probably owing to the swiftness of the current that ran beneath us, when we slung our hammocks among the trees. We could find no place to boil our tea on, so we had some *guarapo fuerte* (cold grog) instead.

27TH.—We went along before a light breeze, and met at sunrise one of the large Apure lanchas, her square canvas towering up high enough for a brig. The river appeared to have gone down about a foot and a half since the 15th. Passed the site of the old settlement of Alla Gracia. The lancha came up with us, the " marenaros" blowing loudly for wind on their horns. Another hard paddle in the afternoon. Made fast to a bush to sleep, above an island of rock called Bonita. No dry spot to cook on. We had cassave, cheese, and grog for supper again.

28TH.—Heavy rain fell before dawn. We lost part of a fine breeze in stopping to dry our things and cook on a coarse granite rock hard by the pueblo of Bonita. After this, we ran past the river Yugape, a good-sized stream, on the south side. The hills of the Cuchivero rose to view. My map was regarded by Ramon, my Indian pilot, with great respect, especially when I read therefrom, and told him of places in his country he little thought I had ever heard of.

We camped on Isla Palmano, noted for its tigres. Ramon did not seem much to relish the situation I had chosen, and much less when we saw fresh tracks of formidable size on the soft soil. However, having charged my gun, I slung my hammock by a good fire. Ramon and Rogers retired to the curiara to sleep. The pelting noise of the fallen yellow hobo plums continued all night, but nothing else disturbed my slumbers.

29TH.—To-day another long view of the river was stretched before us, meandering away into the Apure mouth. I was deeply impressed by the immensity of the river system on which we were voyaging, as these long reaches, with the rounded banks of snowy cumuli, rose above a horizon of water.

We had some caribee for dinner to-day. This fish, though generally no larger than a perch, has such powerful teeth, that it is really dangerous to the bather in waters where they abound. One of them, which I thought dead, actually bit a piece clean off the top of my finger as I was in the act of spitting him for a grill.

30TH.—After a day's rest we crossed over to the north shore, for the first time since quitting the neighbourhood of Angostura. No tea again.

31ST.—Very heavy squalls of wind from the south, accompanied by deluges of rain towards morning. The curiara laboured a good deal. I was obliged to sit up in my place throughout the storm, to fend her off the logs, against which she threatened to stave in her sides. Of course I got a most complete ducking, notwithstanding which I was ready to sink down with sleep.

After sun-rise the weather cleared. We made tea on one of the floating logs, and then proceeded on our way before an excellent wind.

Presently Ramon gave utterance to something between a whoop and a grunt, at the same time there was a considerable splash in the water forward. He declared he had seen a cayman steering straight for the curiara, as if to take a snap at Rogers in the prow.

In the afternoon the sugar-loaf mountain of Caycara was full in sight, rising in an almost perfect pyramid on the south shore. I found that the large grey and white herons did not take much notice of the canoe when sailing rapidly by, at which times I got good shots at them. When we were paddling, they were much more shy. We halted for the night on a long sand island, covered with clumps of the bush so much resembling our willow, and full of sedgy pools, literally swarming with ducks; just the cover for them. Our supper consisted of a large heron I had shot, two macaws, some ducks, and jerked deer-meat.

SEPTEMBER 1ST.—We were glad to continue our halt to dry our clothing. My mackintoshes were nearly worn out, and so torn by the thorny thickets through which we often had to force a passage, that they were now anything but waterproof. Perhaps stout horse-rugs would be better in this climate. They would certainly last longer. We killed plenty of fat ducks, and soon were able to resume our places in the canoes, and started up the Guayana side again. It was very hot, and towards sundown the sky lowered. As we paddled along the mouth of the Cuchivero, I could hear the ominous murmur of the approaching storm in the forest behind. We made a spurt across, and lay-to on the left bank of the stream. Then the storm broke in all its fury,—the lightning flashing, the thunder rolling, in ponderous volumes of sound, from one part of the heavens to the other. I weathered this, as I have many others, nose and knees together, under my blanket.

This Rio Cuchivero is occupied by Indians of the Panare tribe. Their territory extends to the Mato, and other west tributaries of the Caura. We managed to get breakfast after paddling to some distance. Every day, dry land became more visible, now that the river was subsiding. We passed the little grazing village of Taroma, over which immense numbers of the Zamora vultures were wheeling in concentrated circles, after their manner. The bank for a long way this afternoon was the abrupt ledge of a savannah rising from the water like a wall of large stones roughly put together. I shot two fine iguana.

Rogers gave out, and we made fast to a bush at Point Sueño, below Caycara.

2ND.—We put into shore for cooking purposes: there was an abundance of fish,—we caught three pyara in a short time; they resemble salmon in shape, but have two tusks in the lower jaw, like wild hogs. After a calabash bath (for fear of caribees) under the pleasant shade of an overhanging thicket, we paddled up to Caycara. These baths are, to use Ramon's expression, "Muy savaroso," after the fatigue of a journey: the process consists in merely pouring water over the body with a tutuma, or calabash. At Caycara we purchased some papalon (native moulded sugar), rice, cheese, pulpiria, &c., as much of the provisions brought from Angostura were spoilt, from being constantly wet. I had to sacrifice the remainder of my tea from the same reason.

I saw the prefecto on landing. He said he would give me a "pan papero" for Urubana, though I told him it was unnecessary, as I had a through-pass from the Governador. He did not seem to understand me: possibly he did not like to let slip any opportunity for the display of official dignity. We took a dram of rum together, and, as the custom there is, drinking the spirit first, and then swallowing sugar to help it down. During the chat, in such cases not to be escaped, my entertainer spoke, as these people invariably do, before your face (although without exception they call you "Nurienza" behind your back) most flatteringly of "Los Englez y Americanos," and entered on a tirade against the Spaniards of Europe.

3RD.—Under weigh again: we were off Cabenta at mid-day. We fished for dinner, and took a fine sardinato, which is a herring in shape and a salmon in size, of tempting fatness. Towards evening we passed a high sand-cliff, perfectly riddled with the nesting holes of the large red-breasted kingfisher. The chatter of these birds is like the springing of a policeman's rattle, and they vociferously protested against our intrusion on their domestic privacy. We rounded a promontory called Capuchin, which reminded me of Mount Edgecumbe park from some points; but the spaces between the outcropping rocks were much greener, and the trees had a wilder and more eccentric growth.

We camped on a rock for the night. Close by an immense flock of snowy egrets roosted. They were very noisy, as the sun went down in a glorious sky. It may seem odd to the reader, the satisfaction I had in exchanging the following few words with Ramon, after the day's paddle: "Bastante lejo, Ramon?"

To which he would answer, in a tone which an Indian alone uses, "Lego!"

We were driven from the camp by the zancudos. The gnats, or true mosquitos, are called in Venezuela zancudos, whilst those venomous atoms, sand-flies, are called mosquitos.

4TH.—Our next mid-day halt was opposite the mouth of the Apure. I shot some powis, and then had a nap on the rocks. In the morning we had passed along a low muddy shore, much frequented by cayman, then posted on to a hill of big stones heaped one on another, with bushy trees and bright green grass growing in the interstices between: this was the hill Curiquima. We moored the curiara in the cleft of two large rocks, which formed a natural dock for our little craft, and on the summits we spread our blankets. The zancudos were more endurable than before.

5TH.—The Orinoco has in the dry season immense beaches of sand extending from the forest; now, the water reached the trees, and partly inundated the banks. It is wonderful to what a degree the Orinoco preserves its breadth throughout its course, to within the immediate proximity of the cataracts; even here it has a breadth of four or five miles. I had some difficulty in writing my notes on account of the irritation produced by the sand-flies (mosquitos); indeed, I considered them to be the primary cause of a touch of fever I felt in the afternoon. The zamora vultures were very numerous and obtrusive on this part of the river;—

when one is unwell, it is especially unpleasant to
have a mob of these disgusting birds fluttering and
croaking disagreeably in the trees you select for
shelter and rest: sometimes I grew irate, and could
not help sending a bullet from my Snider at one
more intolerably impertinent than the others. If a
bird was struck whilst walking on the ground, he
appeared simply to lie down suddenly on his side;
there was no kicking: the ball drilled a hole through
the body, and continued its way.

It was singular that my Indian, Ramon, could
not hit anything with the rifle, though I never knew
him miss a shot with my double-barrel Holland gun.

It would require the inimitable facility of Ernest
Griset's pencil to do justice to the grotesqueness,
half weird, of a group of zamoras, when skulking
about a camp, dodging behind stones and bushes, or
peering down from the boughs over-head. I was
informed that their chief breeding-places within
Venezuela were among the spurs of the Andes, in-
tersecting the province of Coro. Here they are said
to render some localities quite unendurable by their
odour. The young are downy-white.

6TH.—A fair breeze. I was glad of the chance
to give the steering paddle to Ramon, and to creep
under the shade of the sail, as the calentura was on
me again: the morning was very hot. After the
breeze died away we landed to get something warm
to drink, and found a very damp forest for our
fire. The summit of the Serra was behind us (we
had passed it on the foregoing day); irregular masses

of rock crown it in some directions: it looked like
an ancient fortress, overgrown with trees to such
an extent that the line of the building was hidden.
I gave orders to make fast to a bush in mid-stream
(apparently rather against Ramon's wish), to be
clear of the zancudos for the night, and was aroused
just as I had subsided into a slumber on my gun-
case, by the curiara filling with water from a
sudden squall: we were very nearly swamped, and
I expected every second to feel the canoe go down
under me. The night, to add to our ill-luck, was
pitchy dark, and I could only discern Ramon and
Rogers by the rapid flashes of vivid lightning.
Ramon looked really terrified, and was yelling
out something, which of course I could not distinguish in the din of the tempest. We all baled
away, working with a will, to keep afloat: in the
mean time, I was turning it over in my mind which
would be the nearest way to the shore, in the event
of our having to swim for it. At length we drifted
down to some bushes on a flooded island, where
we braved it out. I never knew a thunder-storm
to remain so long directly over-head.

7TH.—It continued to rain wretchedly for the
greater part of the day, but we reached a conuco
(or clearing), where I determined to stop and make
an endeavour to get dry clothing to lie down in.
Everything in the curiara was drenched; the
mackintosh worn out; and I was afraid lest all our
provisions should spoil that we had in store for
the journey, for the way was yet long.

8TH.—My calentura was much better in the morning; the conuco people had made me a hot decoction of a plant called frigosa. The men did not seem at first very charmed to greet me, but after a little rum was distributed, and some caratero ducks cooked, their hearts were opened, and, re- laxing, they became very obliging. I was regaled with a most refreshing draught of freshly crushed sugar-cane juice. A "blanco" Spaniard arrived, bound for the Cabullare, and an old Indian, who took great interest in my pulse, and wanted to take me home with him to be doctored by his women for a few days; but at evening I felt all right again, with the exception of a slight giddiness, though in the afternoon I had had a bout of ague and fever.

9TH.—Feeling much better in the morning, I made a start, calling on my sympathizing old friend as I passed his place; he again pressed me to remain, but as it was unnecessary, I refused, desiring to push on for Urubana. I took as deep a draught as I could of cane-juice before leaving him. In spite of the "plaga," as they call the pest of mosquitos and zancudos, there are many conucos on this part of the right bank: they are principally of cane, and the most juicy I have seen.

Towards noon, as I felt the fever coming on again, I left the paddle, and went on shore to lie down where it was shady; but becoming worse, I thought it advisable to return overland to the clearing last passed. Whilst Ramon brought the canoe, I managed to control my legs just long enough to stumble into the rancho, where there were two pretty Spanish-looking girls. I did not remember anything more until the fever lessened, and I drank some frigosa the girls had prepared.

10TH.—I had a good night, and was, as usual, better in the morning. There was not much river-view here, as it was divided by a very long island, Isla Cieba, on the other shore, where a tribe of "Indios bravos" (Tortugas) are said to live.

Ramon generally brought in plenty of panji or caratero ducks; but I was always obliged to give him a limited number of cartridges, as he could not resist the temptation of firing away all he had in his pouch.

The fever returned at the same period of each day; I drank hot drinks when shivering, and cold, as soon as the skin moistened. When I was better I took my seat in the curiara, and dropped down to the larger conuco of my old friend. The night threatened rain, so I thought it imperative to attempt a push for Urubana; besides, I was considerably reduced in strength, and could not, therefore, paddle strongly.

14TH.—This day passed without an attack, and I felt nearly well, thanks to the unremitting attention of my kind old Indian friend, whose name was Cumane. Indeed, he only seemed anxious that I should stay with him longer, and expatiated on the richness of the soil of his conuco, the plenty of manatee, and other game. He said he was old, and gently hinted that he had saved much silver (vastante reals).   I think he imagined that probably a young fellow like myself might take a fancy to one of his numerous women-folk, and so stay with him altogether.   One night Ramon awakened me, saying that a lancha of the Rio Negro was passing : I jumped up and hailed the patron, but he declared there was no room on board.   It was a large lancha, containing a good palm-thatched toldo, and was creeping up stream under the strain of twelve blades. I secured an Indian from it, who proved to be a cousin of Ramon's.

We fared sumptuously at our present quarters ; in the meat line, besides fish and turtle, a manatee was harpooned.   I had seen none of the latter on this river until then.   Those familiar with this animal must, I think, pronounce manatee to be the most superlatively delicate and richest of meats. The best description of the great beast of the river is to compare it to a hippopotamus with flippers instead of legs, and a broad tail placed horizontally on the body.

15TH.—I was thankful to feel all right this morning ; so, saying "Adeos" to Cumane, I took my station in the curiara, and pushed once more away up stream.   We passed Caño Civiripe, noted for the number of manatee taken there.   We halted just below Tortuga, for cooking.   In the dry season, this is one of the places at which the Indians assemble to collect turtle eggs, from which they make oil.   There was a grand storm at sundown, but it dispersed before our dinner-time. As it was fine moonlight, I resumed my steering canalite, and we kept on all night.

16TH.—A short sail in the afternoon; a long stretch of the forest on the bank was composed mainly of a small tree with a smooth bark and perpendicular limbs. Travelling in this way, although hardships appear in hosts, excitement stimulates every power of endurance, and if ammunition is plentiful, the physical man is well provided for, as meat offers itself to the sportsman in abundance; and what would be anywhere considered the greatest delicacies, are to be obtained merely for the trouble of sending a shot with a steady aim.

I saw for the first time the curious creature called *mata-mata*, which approximates to the turtle in formation; but the shell is much more rugged, and the head larger, flattened, and covered with toad-like skin. The Indians insist on its being more savoury than the turtle proper; we only took the eggs from the one we caught, as it was old and meagre. It seems a simple, harmless animal, relying chiefly on the great strength of its shell for protection from numerous enemies. The Indian had no difficulty in lifting ours into the canoe from under the bank, where it apparently slumbered; the shell was indented by the teeth of caymans.

Aranca to the right; a fine panorama lay before us, the Serra of Urubana deep blue against the sky, and the hills of the Baraguan Straits beyond. We spent the night on the beautiful sandy shore, some distance below the pueblo, after having cooked on the bank of the savannah.

17TH.—Urubana was originally a mission pueblo of the Ottomac Indians, no descendants of whom are now to be traced about the neighbourhood: some of the tribe are still on the other side of the river. The prefecto was a Spaniard of the higher class, and showed much gentlemanly kindness of demeanour. The pueblo is situated at the foot of a nearly perpendicular hill of rocks, overgrown by trees. Several of the rocks are said to bear inscriptions, but I had no time to examine them. I found the Rio Negro lancha here, and, as it belonged to Castro, the Governador of the country beyond the cataracts, I arranged with the patron to carry myself and effects past the raudales. Although slower, this was safer, and, above all, a drier mode of proceeding than in my little lancha to the required destination.

CHAPTER V.

SEPTEMBER 18TH.—We left Urubana in the evening, and pushed up stream with the combined force of twelve Indian rowers. I despatched Ramon ahead with the curiara.

Above, the Orinoco was much divided by islands. Towards mid-day we had a fine breeze, and passed the hill of rocks called Jovito, which has an enormous slab of bare black rock facing the river. It was a most beautiful night. The Indians, after supping off monkeys, kept on, leaving the Baraguan hills behind. I now travelled much more at ease, and was able to enjoy the scenery. The top of the palm-thatched toldo makes an agreeable lounge. The shore continued to be alternate savannah and low scrubby forest. After passing Juapure, the banks rose to high-water level, and the woods improved.

20TH.—This morning we saw an encampment of Indians of the Paruro tribe, on a sandy spit, in the midst of the river. At noon we crossed to the left bank, which is high, and covered by a long line of plantations, called Santa Barbara: the curious cliffs of Serra Pararuma appeared on one side, and those of Baraguan on the other. We formed quite

F

an imposing force in camp: twenty-four, besides several women who inhabited the interior of the toldo. After dark we continued on past the island of Pararuma, which is famous for the immense number of turtles "that here do congregate" in the dry season, to deposit their eggs, from which the natives obtain a large quantity of oil.

The effect produced by the wild cry of the Indians, uttered from time to time as they laboured, was in singular harmony with the night. It was not unlike the opening notes of some fine, deep-toned hymn; but just as one might fancy it swelling into complete grandeur, it died away abruptly. However, this vocal exercise seemed to have a great effect in stimulating the poor fellows to their monotonous work at the sweeps.

22ND.—Pararuma and Caño Paraguase were now left behind. Had it not been for my gun, we should have been quite on fish fare, as no provisions were to be had; though several times the lancha stopped to get communication with the plantations below.

23RD.—A very wet evening; but in the night it cleared; so we left the sheltering, but close toldo, and had a good supper off some powis Ramon had brought in. We took one meal on the great rounded rocks at the foot of the Caribin raudales, and had our mid-day refection opposite the Cassanare (not to be confounded with the Caniguiare of the Rio Negro and the Orinoco).

On the top of the large domed rock on which

the men were cooking, a great grassy plain was to be seen hazily stretching away towards the long range of hills, which ultimately, when they cross the bed of the river, occasion the cataracts of Atmes. Myriads of fire-flies sparkled like gems low over the surface, seeming to give an undulating motion to the misty plain. The great river Meta falls into the Orinoco on the left shore. The wide plains in the vicinity of this river are inhabited by the dreaded Guahibos, who appear to be perfect Ishmaelites, whose hand is against every man. These bravos seemed to be considerably feared by the mansos (tame Indians) and creoles. Little else but Quahibos's deeds had been talked of for the last few days. These savages possess no canoes, and when they desire to cross a stream, they are said to construct a temporary conveyance for the purpose, of manriche palm leaves.

24TH.—Passing the rocks of San Borja, we came into view of the mountainous country of the Piaroa Indians; like all the old missions of the Upper Orinoco, that of San Borjo has long been abandoned, and the Indians have returned to the mountains and forests.

25TH.—We caught a plentiful supply of a very delicate fish, *palometa*, for which the lines were baited with a bright kind of beetle, and cast in quiet waters among bushes. I always sent Ramon forward to look out for game, as even our stock of manioc had given out. The Zamora vultures were particularly troublesome at this halt. An Indian caught one with a baited fish-hook; plucking out two of the long tail feathers, he stuck them through the nostril cavities of the beak, so as to form a pair of ferocious mustaches. Released with a kick, the unfortunate bird of course endeavoured to rejoin the mob of his companions, which had been gravely looking on; but the more he tried to do so, the more they edged away from him, and at last took themselves off altogether.

26TH.—The afternoon was spent in wending our way up a narrow arm of the river, betwixt forests.

27TH.—A full view of the Atures Hills. Last night we slept within sound of the Great Cataracts, and at daylight the Indians began to strip the lancha of everything for the long portage of some three miles to the pueblo, preparatory to the labour of warping her up the raudales. These men all spoke Spanish, but conversed in their own Baniva among themselves. This is the tongue generally understood by the Indians of the Guainia or Rio Negro de Venezuela, and the Atabapo.

Later in the day I crossed to the pueblo, partly by water, and then over the savannah and Caño Cataniapo, by means of a ricketty canoe, kept for that purpose, moored to the bushes of the bank. The Indians have here, as elsewhere, a strong propensity for always keeping to the old track during their journeys. In the path to the pueblo the track was worn smooth, and well defined across the masses of rock in the portage round the cataracts.

How very ancient it must have been! Probably the Carib war-parties of long ago followed precisely the same tracks over these foot-worn rocks, in their expeditions from the Lower to the Upper Orinoco. The pueblo of Atures is situate on a high savannah, between the little river Cataniapo and the cataracts. In spite of a beautiful site, there hangs about the place an unpleasant air of mortality, perhaps from the association of its name of an extinct race, whose sepulchre is not far off. The four remaining inhabited houses are placed at irregular intervals round the grass-grown square; the intervening spaces indicating where others once stood. Nothing now is left of the old mission, save two copper bells, bearing the date 1740. The general appearance of decay was not diminished by the appearance of the house allotted to me: one side had fallen, and the rafters projecting against the sky had a very skeleton-like look, as the moon rose from behind.

The surrounding savannahs much resemble those of the Caura, but are of a more uneven surface, and more enclosed by rocky hills. These were eminently suited for cattle; but the few herds formerly possessed by the inhabitants have been killed off by an obnoxious species of fly. The poor natives of the village have the aspects of London sweeps, owing to their faces being covered with black spots, that are left after the attacks of the mosquitos (sand-flies).

The Cataniapo is a deep, rapid stream, of most

beautiful sea-green water. It is a delightful spot for bathing, as it is cool and limpid, and free from the dangerous fish and reptiles infesting the Orinoco. Once out of water, however, one has quickly to jump into one's clothes to escape the swarms of mosquitos. The source of the Cataniapo is in the mountain-country of the Piaroa; it falls into the turbid Orinoco in the middle raudales of Atures. At night the roar of waters is very loud in the pueblo, though it is scarcely audible during the day. This change must be caused by the ceasing of the sounds of life, which pass unnoticed in the hours when every one is busy. But when night silences vociferous humanity, the voice of nature is heard plainly. When the sun sinks below the horizon, the chorus of the waters swells into majestic music. The Indians say the raudales sleep during the day.

At Atures, I saw the curious bearded monkey, with hair like that of a negro when carefully parted and brushed down the centre. The natives call these creatures *Capuchins*.

I met Señor Castro; who held sway as Gover- nador of the country above the cataract claimed by Venezuela. He was waiting for his lancha, and received me very well, insisting upon my making use of his own mosquitero, or mosquito curtains, as mine were in the canoe with Ramon and Rogers, who had not yet come up. The Governador's rum flowed freely, and the wretched people of the place were soon in a deplorable condition; the time was

spent in singing drunken songs, improvised in a
cracked falsetto, in honour of the donor.

OCTOBER 2ND.—I traversed the savannah to the
hill of rock, where is the *campo santo* of the
extinct race, the Atures. Part of the savannah
was marked by old cattle tracts, showing where the
kine of the old padres had been wont to pass.
Ramon (who had arrived with Rogers), told
me that there were " vaca" (cattle) still on the
mountain around, but that they were very fierce,
and only approached the open country at
night.

I found the Atures's burial-place to be a hori-
zontal cleft in the sloping side of a hill of rough
granite, under the shelving ledge of rock, where
was to be seen all that remained of the tribe. The
bones of those uppermost had been a good deal
scattered (though originally coffined in a sort of
mapiri, or basket), the rough flakes of rock under
which they had lain having been partly removed.
Some ghastly relics still were intact in Mapiris
of *cocoso* palm-leaf, in which they had been
embalmed. Many of the bones (those, perhaps, once
reposing in the large urns) were still stained with
a red pigment, and fragments of the broken urns
were strewn about. I was surprised to see the
bleached skull of a horse mingling with the human
remains. Might not this animal have been thus
honoured with interment in their sepulchre by the
old Atures, from feelings of awe and veneration
for the first of these quadrupeds seen in their

country, possibly introduced thither by an adventurous Castilian?

Whilst I gazed into the tomb, a beautiful little humming-bird flitted by, and hovered over the white bed of bones. The view from the slope of rock in front of the sepulchre was most striking:—the windings of the Orinoco, until lost finally in the mountains; the great mountain Urubana scarcely reared its crest above the heavy vapours concealing the lands of the Guahibo bravos. On the other hand stretched the savannah of Atures, its rocky surface hidden under green grasses, with belts of the stately manriche palm, or clumps of Chapparo trees and bushes in every hollow where a sufficient soil had accumulated, and beyond the mountains of the Piaroa Indians. Ramon said that these Piaroa were at feud with the dreaded Guahibos. The situation is singularly well chosen. When I stood there, with the great rock-sepulchre behind me, I could picture vividly how the procession of the Indians bringing their dead for burial, must have approached and passed up the inclined plane of granite-rock in front.

On the Atures savannahs I noticed many pretty birds; one peculiarly so, seemed to combine in its habits those of a lark, a thrush, and a rail.

3RD.—We left the beautiful, but insalubrious savannahs of Atures, which are rendered almost uninhabitable by the plague of mosquitos (sand-flies) and zancudors. That night we left the loftiest of the surrounding mountains, Uriana, away on the left bank, and Serra Luripana, with its sepulchre, a short distance from the right. Many of the Guayana tribes still lay their dead among rocks, after the manner of the old Atures.

With comparative ease the lancha was wafted up the little raudales of Garcita and other small falls. Here the rocks do not bar the whole river, as do the stupendous masses of Atures and Maypures. There were many balsam capivi trees on the banks, but they bore the marks of tappings. In the vicinity of the great cataracts the sky is almost always overcast the year round, and in the driest season it rains there every day.

In the mountainous country beyond the savannahs there are two kinds of deer, the one small, but the other large, with branching antlers. The lancha rounded the hill-island, and lay-to at the puerto of the path leading to the pueblo of Maypures. The night was most serenely beautiful; but the patron somewhat spoilt the effect of the imposing roar of the raudales by screeching some Venezuelan ditty in a shrill falsetto. It is singular that these people endeavour to render their voices as ridiculously effeminate as possible whenever they attempt to sing!

Hearing that vampire-bats were very numerous, I took the precaution of covering my feet when I slept. In the afternoon I walked with Señor Castro across a savannah resembling that of Atures, but even more lovely, though it was infested with sand-flies. The savannah streams of this district are beautiful. We passed over one on a bridge of felled trees: it had a spring of capital water welling up from a funnel-shaped cone of white sand at the bottom.

There were in the pueblo only four or five men of the same mixed class as those of Atures, chiefly from the Cassanare district, I believe. The Indians of the Maypure race had long emigrated in a body for some less plagued region at the head of the Guainia—at least, so the Indian patron asserted. The Piaroa stay in their altitudes, and only descend to the savannahs of the troubled bed of the river to fish and to procure a few necessaries. In manners they are very mild. They possess fine çonucos, or plantations of cassave (manioc). The indifferently-cured tobacco used by the people is of their own cultivation. The native Indians have also retired far off, and there are a few mansos Guahibos on the left bank. The patron assured me that the Macos of the upper Orinoco are a distinct nation to the Piaroa, and speak another language. The scenery in the neighbourhood of the two great raudales, or cataracts, is the most varied and beautiful I have anywhere seen. There are the open savannahs, here and there broken up with those magnificent masses of rock which give this portion of the Orinoco so marked a character, contrasted with the level belts of the primeval woods which border the river, and the slopes of the hills and mountains that are not sheer rocks. The Orinoco itself winds away in the noblest proportions, and is studded with rocky

islands and islets of every shape and aspect. These
are rendered gem-like by tufts of palms and other
vegetation, fostered by the exhalations from the
warm, foaming waters of the cataracts. Their vast
extent and variety is one of the chief charms of these
raudales; there is no abruptly high fall, but they
are formed by steps or terraces of black rock crossing
the river's bed, over and between which tumble
great masses of snowy foam, with an incessant roar.
Between these saltos there are reaches where the
river flows in quiet depth, and is so shut in by rocks
and islands, that nothing but the never-dying voice
of the waters indicates that the canoe is still within
the raudales. One evenly-drawn line on the water-
worn boulders shows the rise and fall of the tide.
As a general distinction between the two cataracts,
it may be said that the slope causing that of Atures
is the longest, and that of Maypures the most abrupt.

The Indians believe that some of the saltos are
more easily passed in the dry season, while others
are less difficult when the river is deepest. I was
introduced to the padre of San Fernando de Atabapo, at one of the houses of the pueblo of Maypures.
In addition to his spiritual calling, he is one of the
most active traders of the country beyond the raudales. He descends to Angostura annually in his
lancha, with his rubber, balsam, and cassave, returning with a fresh supply of commodities of trade.
This energetic priest is, I believe, a Genoese by
birth, and entered this part of the world by way of
the Rio Negro, coming down Rio Madeira from the

Republic of Bolivia. When I first saw him he was sitting in one of the chinchoras (hammocks) of the house, where he looked odd in his monk's habit. Afterwards I knew a good deal of him at Maroa, and found him, on better acquaintance, to be a pleasant old fellow, and not an excessive dogmatist.

Nothing was to be had from the slothful people of this miserable pueblo but a little poorly-made cassave. Drunken carousals continued without intermission (El Governador being chief instigator): the noise and annoyance was most disgusting to a person who was obliged to be an involuntary spectator. Castro at length reduced himself to such a pitch of nervous excitement, approaching to phrenzy, that I thought it advisable to give him an opiate, which had the desired effect. Our fare here was a little fish with cassave, about once in the day; this, added to the distressing irritation of sand-flies by day and zancudos by night, made the place unendurable. Castro, being recovered, made a start for his own pueblo, and I accompanied him. Wishing, however, to make a forced march, El Governador plied his men with so much rum that the consequence was, after we passed Punta Raton, and were near Rio Vichada, it was discovered that we were making little progress, as three paddles had been lost during the night. The head of the curiara was accordingly ordered about for Maypures again. To my suggestion, that it was better to keep on now, than to lose so much ground, Castro replied that the governor of the country could not possibly appear at San Fernando in such a plight, with a friend.

Very fine view from the top of a rock sloping up the savannah where we halted. A broad belt of tall manriche palms looked like a stately avenue leading to the two hills of Maypures. I could not help thinking that if the comparatively insignificant palms of the East suggested to the ancient Greeks their beautiful columns and capitals, what if they had seen some of the same trees in the tropical West!

We attempted a fresh start at sundown, but it came to naught; for the men, not having been able to satisfy their hunger enough in the curiara, remained too long in the pueblo when summoned. I was sorry to see Castro bend his bright toledo in thrashing the first offender that appeared. Several of the others took to their heels, and were not afterwards forthcoming at all.

On the banks of the Orinoco, above Maypures, the palms most noticeable are the zagua wine palm, and the cacarito, as is the coroso, lower down in the islands of the raudales. The Macanilla, a thorny palm, here attains great height, and the graceful manac again becomes plentiful: the latter has a wide range. I saw it at the delta of the Orinoco on the Upper Caura, but on the Upper Orinoco it tapers to the greatest height; on the Amazon it is called the assai. The zagua is, indeed, beautiful at all times, but the more so when some point in the river enables it to rear its stately fronds above the lower woods, thus obtaining an azure field for the display of its serrated leaves, drooping over towards the outer side. It resembles the cocoa palm, but the leaves shoot out from the stem, edge upwards, and this peculiarity gives them a more graceful curve. I neglected the opportunity of measuring the leaves of this palm, but they appeared of enormous size.

This morning we met a Piaroa curiara. As it had cassave on board, our canoe immediately put alongside, and the two men (and the cassave) were pressed for San Fernando. Although evidently unwilling, the good-looking matron was left to paddle the curiara home, with the youngsters, goods and chattels, all their little plans upset for that day. It is a wonder that these simple people do not even more seclude themselves in their mountain forests, as this liability to forced service must be very distasteful to them.

We put on shore in the morning to cook, and when it was time to start, the two Piaroa had made themselves scarce. No one had seen them go, but they were nowhere to be found.

The woods greatly improved in appearance as we went on. At noon we were opposite the mouth of the Vichada, which is occupied by Guahibos, who, less warlike than those of the race lower down, have plantations on its banks. The Guahibos I afterwards met with at San Fernando were taller in person, and much bolder in bearing, than the Piaroa. The heat was considerable, and was followed by a heavy storm towards sundown. We next reached the Zama, famous even here for mosquitoes (sand-flies). It is the first of the system of Aguas Negras, black-water rivers, rising westward of the Upper Orinoco. Owing to the high forest we were skirting, we saw nothing of the eastern mountains since we left the neighbourhood of Maypures.

The Governador Castro was an excitable casador (hunter). We went on shore, but shot nothing save *pava*, a kind of game, differing from the panji in smallness of size; the plumage, unlike the *guacharaca* of the lower river, is mostly black, with a white crest. I saw no more of the pretty little grey finches with crimson heads, so common on the banks of the Orinoco lower down. Instead of that species, there was a little grey or drab fellow, which flitted from bush to bush for long distances before the canal.

There had been no general sleeping since we left Maypures: this was the fourth wakeful night for some of the men, and, despite their efforts to the contrary, slumber would, from time to time, overtake them. The Governador gave them freely of a demijohn of Malaga he had with him; and it was impossible to refrain from laughing, as one or other of the dark mariners would drop into a somnambulistic state, in which they kept time to the chanting marvellously with their paddles, although the stroke might be taken in air, and wide of the water. One of the number, who was quite as sleepy, and a little more drunk than the rest, again lost his paddle;

looking round with a stupid grin, for a moment he seized that of his nearest neighbour, and went on as before. All was quiet for a few strokes, when it seemed to occur to the original owner of the paddle that he was lax in his duty, so after a scrimmage he got it back; the odd man gradually subsiding under the thwarts of the canoe.

OCTOBER 9TH.—At daylight this morning we discovered a surface of deep coffee-coloured water, even darker than the Caura, and the numerous cocoa-palms that rise above the thatched houses of San Fernando de Atabapo. Señor el Governador received me most kindly at his house, and installed me in an empty one next door. Amongst his friends there was a most enterprising young Venezuelano Spaniard, Andreas Level, who told me much that was interesting concerning the unknown regions lying between the upper affluents of the Orinoco; boasted of having gone three days beyond the supposed impassable raudales of the Maguaca; and said that he had seen cut into the side of a rock an inscription: " Raudales de los Guaharibos." He described these Indians as being as white as myself, but with red hair. It is strange how these reports coincided with those of the Blanco Guatusos of Rio Frio, in Nicaragua, and other parts of Central America. He was much aided in his expeditions by the Arquiri, and had espoused the daughter of one of the chief men among these Indians; so, in consequence, he enjoyed nearly a monopoly of their good offices, with the produce of their country, consisting mainly of balsam capivi, also finely made chinchoras (hammocks), and other articles that obtain a ready sale at Angostura.

The inhabitants of San Fernando, with the usual improvidence of the Spanish creoles, rely chiefly on these Marquiritare, and the Poyñaves of the Quirida, for their supply of the bread-stuff, *manioco*, for they have not a single plantation near the village. Cacao of delicious flavour grows wild on the bank of the Guaviare opposite, yet the people are too lazy to secure more than small quantities for their own use, although their regular avocation appears to be visiting from house to house, chatting throughout the day. The doors and windows of the domicile I occupied were riddled in various places by bullet-holes,—intimating that even here, in the Ravo de. Venezuela (the Tail of the Country, as they are fond of calling the Rio Negro district), there appear to have been the émeutes, faction fights, common to these republics. San Fernando is one of the very few villages, originally Indian missions, which has since grown in size and population. It probably owes this exemption from decay to its admirable position near the junction of four large rivers,—the Guaviare, Quirida, Atabapo, and Orinoco. The climate is very heavy and close, from a want of breeze; but when the sky is overcast, and one of the frequent torrents of rain falls, I have shivered under my blanket, from the sudden alteration in the atmosphere. Coco-palms (the palm of the sea-shore) abound, though far from the salt sea spray in which

G

they delight; they seem to be very healthy, and have quite usurped the place of the *pijila*, few of which are visible about the village houses. A tree here has a very curious growth, with a stem like the palm; its leaves only cluster in a mop-like head at the top.

The heavy atmosphere and fish diet of San Fernando reduced me soon to a state of debility, which was not exactly illness, but it made me long to be once again in my little curiara on the broad breezy river, where I should, with the assistance of my gun, be able to procure, and enjoy in my camps, a simmering pot of panji. Anything rather than this "fish, fish, fish" of the villages!

My original intention had been to gain the Rio Negro by the route of the Atabapo and the Pissuchin portage; but as my funds were getting uncomfortably low, I determined to recruit them somewhat ere attempting a descent into Brazil. In prosecution of this scheme, I got two Indian peons, and two boys from the Governador, to search the banks of the Orinoco, between the old missions of Santa Barbara and Esmeralda, for ciringa, or india-rubber *(Siphonia elastica)*. At San Fernando I noticed many pretty little birds which I had not seen lower down; among others, a very lively, handsome finch, of a deep rich russet colour, that was black against the sky. Sky-blue finches, common at Angostura and Pará, were flitting about here in the orange trees and guava bushes.

CHAPTER VI.

October 24th.—Ramon and I started this
morning in my old curiara, by moonlight, for a
preliminary exploration of the forests of the
Upper Orinoco. The air was heavy with the odour
of the flowers of the water-loving *gica* tree, when
the sun rose over a rolling bank of mist. We
travelled half the day, and the Serra Sipapo came
into view from our halt on the left bank, on a shady
point of dry forest. These mountains have a very
fantastic outline,· and are said to be a continuation
of the hills of the cataracts. We caught a sufficiency
of fish, but were much pestered by a miniature
kind of stingless bee, which swarmed on any
exposed skin, evidently intent upon lapping and
extracting the salt from it.

26th.—Finding plenty of panji, we camped on
a shady island.

28th.—Above the junction of the Guaviare, the
Orinoco diminishes perceptibly in breadth; the
banks are generally lower towards each point where
the river bends, and afford well-sheltered camping-
grounds. I think nothing strikes the novice in the
South American forest foliage so much as the re-
petition of tongue-shaped leaves, more or less varied

and modified. As a rule, with the larger forest trees, the foliage is rather scanty, as if nature here made provision for the immensity and variety of the creeping plants and parasites.

31st.—Passed half the day in the curiara; there were several rather bad runs above the old mission of Santa Barbara.

Nothing now remains of the mission, situate, in times past, on the border of a fine savannah, than a wooden cross, and the posts of some houses overgrown with guava bushes. We looked in vain for orange trees amongst the thick scrub; so embarked again, and camped above the Guachiriu rocks. A great many sand-flies; from below the rapids of Santa Barbara to the mouth of the Guaviare, there are very few.

November 1st.—We passed up a bend in the river, much blocked by islets: on the right bank is the delta of the large Rio Ventuare. Later, we left the main stream, and followed the windings of a lagoon of very clear, though dark water, in search of something to eat with our manioco. Gazing down through the clear coffee-brown water, it was curious to see the submerged foliage of the trees, branches, flowers, and fruit, waving about in the current; shoals of lively little fish occupying places usually assigned to birds. After losing many hooks by the caribee's teeth, I caught several good-sized specimens for our larder, and Ramon shot a powis. Often when I was shooting ducks in the forest, I have lost them provokingly, from their falling into the water, and being instantly snapped up and devoured by the voracious caribee fish.

2ND. — To-day we started early, and after paddling for a long stretch, we stopped at noon, on a rock on the edge of a pretty savannah, that re-called somewhat part of the English Downs; but here were no signs of men, save the remains of an old camp-fire of some Marquiritare passing to their homes, on the Ventuare or the Conuconumo. One of the pretty little river terns (a youngster) alighted on the rocks whilst we were cooking. I could not resist my predilection for pets, so I baited a fish-hook with a piece of fish, and secured him.

3RD.—Fine view of the Serra Yapacani, a great isolated mountain-mass, which appeared this morning like an immense bar across the end of a bend in the river. Many San Fernando people were at work in this neighbourhood, collecting the *goma*, or india-rubber. For the last few days, the solitary and singular-looking mountain Yapacani, almost always capped by a sombrero, was in sight. I had imagined the hidden summit to be peaked, but on one occasion the cloud rolled from it, and I perceived clearly it terminated in a long abrupt line of ridge. I made a sketch of it.

4TH.—We were in camp towards dawn this morning; and I was unpleasantly aroused by a rustle in the dead leaves near me, and presently the unmistakable "whirr" of a rattlesnake sounded, as the reptile passed round the base of the tree to which the head-cord of my hammock was made fast. The moon had gone down, and the fire was burning low, therefore it was very dark. I cautiously reached a stick from the fire, which sent the intruder a little further off. The light of a brand discovered the "coscabel" erect, rattling furiously at the now blazing fire, and a well-directed charge of No. 3 shot settled the matter. On measuring the snake, when it was daylight, he proved to be 6 feet 6 inches long, and bulky in proportion. I was surprised (and so was Ramon) to find that he had no rattle, and so concluded the loud noise that awakened me must have been produced by rubbing his rough tail against the rasp-like scales on his back. The skin was more handsomely marked than is usual with this kind of snake, being mottled on the back, more after the fashion of a water-snake. The upper jaw was armed with four long fangs (two on either side), that folded back against the roof of the mouth, when not required for use.

5TH.—I came upon the camp of Señor Hernandez, a native of the province of Guarico, who, like the rest of the creole whites, had fled hither into the forest of the Ravo de Venezuela, to be out of the way of the revolutionary troubles of the more populous part of the Republic. He was working rubber in the woods of an island opposite Serra Yapacani, and received me very kindly, as I negotiated with him for two quintals of goma. We had rather short fare at this camp, only one meal at evening, when the pescadores (fishermen) came in. But the people seemed to look on this as a matter of course, and contented themselves by munching a handful or so of dry manioco, or swallowing it, stirred up in a *tutuma* or calabash of water. But this fasting by no means suited our English stomachs. I little thought then, that in a few more weeks, on the Upper Orinoco, I should come to drink *incuta* (manioco water) with any of the Indians, and even prefer it to other beverages, when paddling and exposed to the sun. I remember when I first attempted to swallow it, the hard grains of manioco so tickled my throat as to give me a violent fit of coughing.

Señor Hernandez seemed to be living in hope of better things when the dry season should fairly set in. The thunder was continuous; the storm seemed to converge round and round the isolated Mount of Yapacani.

11TH.—As the men promised me by Señor Castro were not forthcoming, I posted down stream to San Fernando, with Ramon, to look them up, passing the rocks and rapids of Santa Barbara while it was light, and taking a sleep while the curiara drifted down with the current in mid-stream, as I knew there were no rocks in dangerous places below. I well remember the first time I steered my canoe down a rapid. I had put her head for the smoothest place in the inclined plane of water before me, and, having done so, was rather startled by the ominous "whoop" given by the Indian at the prow; but did not fail to aid him with his paddle, and so we shot down in the jumble of broken water. When we got below I looked back, and at once saw the error I had made, in the selection of what I thought an easy channel; for I perceived, under the smooth, raised water, the black rock not visible from above, whilst on either side ran the really deep water, its surface disguised by formidable eddies of foam.

12TH.—San Fernando at noon. We found that the whole of the inhabitants had been seized by a kind of mania for " goma," and were gone "al monto" in search of it. The idea appeared to have struck them that this really must be a good thing, if an Englishman like myself, coming from so far, desired to go in for it. I purchased a larger and slighter canoe, called a " casco," and got back to Yapacani on the morning of the 20th.

22ND.—Again up stream with all my goods and chattels, bound for the Ciringa districts, which were some days' journey further on. At starting, the men called out, as is invariably their custom,—

" Vamos con Dios, patron !"
To which the patron replies—

" Y con la vergen."

Above Yapacani the river is much divided by islands.

23RD.—On the evening of this second day's paddle, we arrived at the large island Puruna-minare, having passed the mouth of a considerable stream of the same name, on the right bank. I stopped at the rancho of Señor Lanches, a native of

Barquisimento (since made Governador, in the room
of Castro, who was kicked out of office). Señor
Lanches was working rubber on the island.
The night before had been stormy, but to-night
a perfect tornado swept down the river, and I
became apprehensive for the manioco provision in
the canoes, fearing the damp would spoil it.

24TH.—Continued our journey, and had fine
moonlight nights. As soon as the sun went down,
and the mosquitos (sand-flies) disappeared for the
time, I used to strip off my shirt, and paddle on in
my "pijamas," leading the way in the long, light
canoe, with the two boys, followed by Ramon, in
the old curiara, with two men. The moon was so
bright, that it was easy to note the indentations
of the bank, and, at the same time, to avoid running
into rocks, and the overhanging trees and bushes.
Ramon could see the white skin of my shoulders a
long distance off as a guide. At these times I often
shot remarkably fine panji, much more so than
any seen during the day, as they were sitting
drowsily in the moonlight on some spray under
which we happened to pass.

27TH.—We paddled on, following the left bank
through a heavy fall of rain in the morning.
Ramon (who as usual was steering the old curiara),
was obliged to give up on account of a most severe
fever chill. We arrived at the mouth of a small
river, called Caricia, or Chirari. As this was about
the neighbourhood I purposed to work india-
rubber during the drier weather, I camped; and

after seeing Ramon properly attended to, I at once despatched the men into the forest, while I paddled the casco up the creek in order to take them in further up. In a short time they returned with their notched sticks, indicating fifty-seven trees seen in the small space of forest they had traversed. I felt satisfied with this intelligence, but next day went on as far as the next creek's mouth above. I discovered here that the Orinoco, instead of receiving, gave off water, which, after describing a semicircle, and blending with the water of two streams, Aguas Negras, fell into the main river by the mouth I had first entered, thus rendering the piece of land I had determined to work for rubber an island. I had noticed fine ciringa trees on the Orinoco bank all the way. The forest on the two sides of this stream presented a marked contrast: the black water following one bank, and that of the Orinoco the other. The forest on the bank occupied by the white water, contained the ciringa or india-rubber trees, the manac palm, and other trees, in striking contrast with the opposite side, which had neither ciringa nor manac, but an abundance of the Chiquichiqui palm—the Piassava of the Rio Negro.

DECEMBER 1ST.—I determined to put up my rancho for the season's work on a well-drained bluff, which abutted above the very dark, clear water of the first and smallest of the two streams already mentioned. This branch creek flowed out of some large lagoons away to the west. It was strange to see the *toninas*, or river-porpoises, disporting them-to face page 90

MY RANCHO, DURING THE INDIA RUBBER SEASON.
CARICIA, UPPER ORINOCO.

selves in this little creek in the very core of the continent.

Having thus fixed on working quarters, I sent Rogers and two men in one of the canoes to the plantations of the Marquiritare on the Conuconumo, to negotiate for the necessary manioco. Here, in my little creek, I felt indeed shut out from the rest of the world. After passing within the mouth, and taking a few turns, all trace was lost of nearness even to the unpeopled reaches of the Orinoco, so completely is this river enclosed by the forest. Caricia, even more than most parts of the Upper Orinoco country, is dreadfully plagued by mosquitos (sand-flies). The true mosquitos are not as troublesome at night as they are in places lower down. I used to watch the cold shadows of night gradually creep up from the water on the opposite side of the creek, and when the topmost boughs of the forest trees were alone tipped with golden light, I had the fires built for supper. It was the only time of peace throughout the day. Then the long-drawn note of the Gallina del Monte sounded somewhat sadly from different directions of the forest. The eggs of this bird were plentiful and very acceptable. They are laid on the bare ground among the roots of trees, and are of a beautiful blue colour, though like those of a partridge in shape. Another singular note heard from the forest is made by a kind of rail. These birds appeared to assemble in numbers in swamps, especially when it threatened rain, for then they created such a clatter as to make the very woods resound. The ibis of the Orinoco is the most restless of birds; its cry of " coro-coro" is heard at the first streak of dawn (whence its native name); and with quick flapping wings it flies down into the silent creek. Even when the moon rose late and all was still, save the occasional voice of an owl or *rana* (tree-frog), the sound of " coro-coro" would ascend from the tree tops, as the bird took a sudden restless flight. The Indians say " coro-coro" sleeps least of all birds. A small pigeon's note is also often heard in the forest; it is very like the Spanish words " falta-poco" clearly intoned.

The constant irritation from the bite of the mosquitos at length caused my hands and feet to swell, and become inflamed, and, after a time, to break out into distressingly ulcerated patches on the knuckles and backs of the hands. My feet especially were so inflamed, that I was confined to my hammock for some days, whilst Ramon and the two boys were putting up the lodge. The last capping having been given to a substantial roof of palm leaves (those of the all-serviceable chiquichiqui), Ramon and I went to work for the first time on the india-rubber trees. My plan was to cut a path along the Orinoco coast, and another along the creek, and then to intersect the triangle of forest enclosed betwixt them. We found the forest dry and good for work; and, at the beginning of my task, on the very first day, I cleared sixteen trees with the assistance of the two boys, Ramon cutting

the path with machete. My forces all told were but
small, viz., Ramon, who had grown accustomed to
my ways,—I had dubbed him patron (head man),
although he and Rogers did not get on well together;
Mateo, a queer weazened-looking old fellow, who,
whenever he glanced at me, assumed a most insinu-
ating grin, making me feel as if my own features
were involuntarily taking the same expression:
this old boy was the worker of the party, and did
everything with all his might; Benacio, a stolid
old man, with no particular attribute to mark him
from the others, except that he ate more than his
comrades, and was generally in slow marching
order when in requisition; finally, the two boys,
Narciso and Manuel—the former, though as big
again as the latter, was decidedly stupid; he did
not seem to comprehend Spanish very well, and so
it was "trabajaso" (as Ramon would say) with
him, in my rather limited vocabulary of that language. Manuel was a very bright little fellow,
somewhat approximating in character to a London
street boy. I made him a sort of body servant, but
I afterwards found him very roguish, and more
given to pilfering than any of the others, which is
saying a great deal, as almost all of these peons, or
reduced Indians of the Rio Negro and Atabapo, are
great thieves. They rarely take things of much
value, but are systematic in stealing whenever they
have the opportunity,—apparently for the mere love
of the thing, as I have often known them to purloin
articles that were perfectly useless to themselves.

On the 13th, Rogers returned with only twenty mapiri of manioco. I continued cleaning the trees daily in the forest, and hoped to have 1,000 ready for tapping in the ensuing month. One of the chief features of the forest is the variety and immense number of bush-ropes, "bejucas," forming a sort of natural cordage; they are of every size, and bind the top branches of the trees together, winding round the trunks, and coiling themselves on the ground in endless snake-like contortions. In some places they caused the men much trouble, in cutting the paths with their machetes connecting the ciringa trees. Amongst the species I noticed one kind, the section of which, when cut, tantalisingly resembled the roly-poly jam pudding of home days. Sometimes, during the time for rest, I would sit down and look up into the leafy arches above, and, as I gazed, become lost in the wonderful beauty of that upper system—a world of life complete within itself. This is the abode of strangely plumaged birds and elvish little *ti-ti* monkeys, which never descend to the dark, damp soil throughout their lives, but sing and gambol in the aërial gardens of dainty ferns and sweet-smelling orchids, for every tree supports an infinite variety of plant life. All above overhead seemed the very exuberance of animal and vegetable existence, and below, its contrast—decay and darkness. Here and there was a mass of orchid, carried from above by the fall of some withered branch, sickening into pallor, thrust out from the vitalizing light and air.

When the fruit of the ciringa *(Siphonia elastica)* approaches maturity, it is first visited by a flock of parrots, and then by the harshly screaming flocks of the yellow macaw. These birds are most wasteful feeders, the ground beneath the trees becoming speedily strewn with untouched fruit as well as the shell of the nuts.

There are many kinds of monkeys in the neighbourhood, from the large red *originato*, which roars hoarsely (making a far more formidable noise than the tiger) at any change in the weather, to the pretty little ti-ti. A troop of the latter is one of the merriest sights imaginable, as they bound with wonderful agility from bough to bough, leaving no leaf within reach unsearched for its lurking insects : they are especially fond of the leaf-winged locust. The little creatures look truly elf-like as they peer down at you from behind a screen of foliage to get a clear view of so unwonted a presence, before they scamper off and away through the clustering branches over-head. The whistles of the monkeys, greatly resembling the notes of some bird, are heard from different parts of the forest, as they answer one another. The *arizualos*, unlike the deep brown and black monkey of Central America and the lower Amazon, are a rusty-red species : they are equally surly, and give vent to their feelings in the same monstrous volume of roaring sound as the originato. Snakes were very numerous, and of great variety in form and colour. Ramon had no names for many that we saw : they generally managed to glide quickly out of the path, and so escaped me. There was one pretty little reptile more impudent than his brethren, and less inclined to get out of the way. It was of a beautiful green; the Indians call it *loro* (parrot), and Ramon said it was very savage and venomous—"muy bravo."

The orchideous vine-vanilla was common in the forest, but it seemed rarely to bear fruit; and when it did so, the pungent luscious aroma was to be perceived from a distance. There were many tigers, as was evinced by the numberless tracks in the woods: the Indians were sometimes scared from their work by this terrible footprint, but I never personally encountered a tiger here. Occasionally I saw a freshly disgorged fish, in the path between the ciringa trees.

We had some heavy rains at night, and no matter where I shifted my bed, the water would drip through the roof on my head, putting anything like comfortable sleep out of the question. I came to the conclusion that the tilt of the roof was not sufficiently steep, and so ordered the men to cut some saplings, having a fork at the top: these we placed as temporary props under the chief beams. We then took down the supporting posts, and, having cut them down to the right length, replaced them, and gradually removed the temporary props, until the roof rested on the reduced posts, at a more acute angle. This had the desired effect, and the rain no more came in upon us at night.

Two days before Christmas I sent Ramon with Mateo to hunt up a wild hog or a deer for the festive occasion; however, they returned with nothing more than a panji, so the men had to fall back upon fish, which the waters of the creek rarely failed to yield in abundance. Christmas Day was spent in the rancho: in the morning the sand-flies seemed rather less troublesome than usual, but in the afternoon they appeared in swarms, and in the evening Rogers had one of his fever-chills.

There was small chance here of over-sleeping the dawn, for with it came the mosquitos, and they do not desist from annoying until dark. My feet and hands again became very sore and inflamed, from the constant irritation of these plagues. Daily wishing for night is not a very satisfactory way of living. I did not suffer quite so much at this season, when working.

Christmas Day past, I despatched Ramon to gather a supply of the old nut-shells of the cucurito palm, used in smoking the rubber. The day being fine, I commenced tapping with part of the people; the others continued to clean more trees to be in readiness.

On the 2nd of January, 1870, the creek underwent a change in appearance. The current had long ceased to flow, and a sudden rise of the Orinoco had caused as sudden an influx of its white water. Latterly we had been much troubled by a large-headed worm (guoan) appearing beneath the skin. The Indians said it was produced by the Zancudos

H

Colorado (the red mosquito), which had become very numerous in the woods. I think the Indians right in considering these to be the larvæ of a gnat. Those Ramon extracted from my back had precisely the shape of the wriggling things to be seen in most rain-water, enlarged, however, by the fostering heat of the flesh in which they were embedded. They also appear to breathe through their tails, as the head is buried, whilst the pointed tail-end approaches the surface of the skin. Their presence is not noticed except when they feed (at least I presume so, from my own sensations). The first time I felt them, I could not imagine what on earth was the matter with me: it seemed as if some one was making a succession of thrusts into my side with a red-hot needle. The operation of extracting the insects is tedious and painful: they are first killed by the fresh milk from the india-rubber tree, or tobacco juice, applied to the red spot indicating their lodgings. This district is plagued by the mosquitos beyond any other spot I visited; added to these are biting ants, chivacoas, niguas, wasps, &c.

8TH.—I had tapped the first hundred trees, but the yield was very small, which disappointment I attributed to their being loaded with green fruit. On Sunday I often paid a visit to a friendly creole, called Merced Gil, who had followed me from San Fernando, and had established a rancho a few hours' paddle up the river, working the woods for ciringaro, near the Serra Caricia. He stated his willingness to supply me with a casco (a large canoe with the extremities squared above the water), and complained of being on "short commons," having nothing in his rancho but the salted flesh of a wild cat, to obtain which delicacy he had loaded his French gun with ball, and had, in consequence, blown a hole in one of the barrels. I saw the skin of the beast hanging out to dry; it was of uniform grey, thereby differing from the usual forest cats of South America. I was better off for fare at my creek, with its fish and occasional fowl.

Balenton is the largest and most powerful fish of the Orinoco. Ramon hooked one that towed the canoe a considerable distance, and he was ultimately obliged to cast off, as the line was too short. The balenton frequently leaps from the water when in pursuit of its prey of smaller fish, and in its heavy flop back into its native element raises a cloud of spray, seen from afar.

We should have lived well here, but that my ammunition was fast giving out, so that we were unablé to kill much game. There were also plenty of the *mono-chocote* (a monkey with long red hair and a short tail) to be shot on the shores of the succession of lagoons, or lakes of black water, opening out from the creek above. This kind of monkey is particularly esteemed by the Indians. It may be said that of all generally known meats monkey most resembles hare, being dark and stringy. At a place betwixt these lagoons of Agua Negra, I saw, on some rocks that lay exposed when the weather was driest, concentric lines cut deeply into the surface. What does this peculiarity indicate? Can it be that the swamps and lagoons lying between this and the Atabapo were once inhabitable, and inhabited?

From the furthest off I visited was to be seen a high hill or mountain, rising over the swamp growth to the south-west. I believe it is possible, during the height of the wet season, to pass, by way of a network of lagoons and connecting creeks, round the base of this hill to the Guainia or Upper Rio Negro, or to communicate with the head stream of the Atabapo.

As the weather became drier, another plague increased upon us, niguas (jiggers). My neighbour, Merced Gil, told me that in his eight years' experience of the Upper Orinoco, he never knew the waters so high as they were this season. There had been a slight subsidence, but now the water rose again. Every day and night we had heavy rain and an overcast sky. Turtle was generally plentiful, except at this time: we only caught a few of the small species, "terekya," and fish became scarcer. The flies were most troublesome,—we could hardly preserve anything from their contamination. Even if the men left a few small fish in the curiaras, in a very short while they would be completely lifted up by such masses of eggs as to resemble honeycomb. It would have seemed incredible had we not seen them. I was obliged to cover the troughs in which I put the liquid rubber, to prevent its becoming embellished with self-immolated blue-bottles. The bite of the scorpion of the Orinoco is not so painful as I had anticipated, nor does it occasion any after bad effects. When I was stung, the smarting and accompanying feeling of numbness was not so great as that caused by the sting of the forest wasp. Cockroaches, the irrepressible pest of some parts of Tropical America, are not so numerous here. A kind of kite was a great nuisance: besides the disagreeable squall of these birds, they often swooped down and helped themselves to the salted fish, as it hung out to dry, and sometimes succeeded in bearing off large pieces in their claws, in spite of shouts and sticks. I have seen them rising from the ground with a long snake dangling from their talons. One day I discovered a new depredator, in the shape of a magnificent rey de las zamoras (*Sarcoramphus papa*), or *urubu-tinga*, the king of the vultures; but he rose majestically, and soared away before I could get out my rifle. He was very large, and in beautiful plumage, but I was getting too hard up in ammunition to use my shot-gun to secure him. I consoled myself, however, by thinking of the pangs of thirst he would suffer after such a gorge of salt fish.

Having looked up all the ciringa trees within the triangle of my paths, I continued to tap them daily, as the weather permitted, though the result was not very satisfactory.

31ST.—During the last week we had a visit from a party of Marquiritare, on their passage home to the Conuconumo. I thought this was a good opportunity of sending Rogers to procure more manioco; he was of no use to me here.

These Marquiritare are the most numerous and important tribe at present on the Upper Orinoco. They live chiefly on the banks of the Conuconumo, Paramo, and other tributaries on the right bank, and are much fairer in complexion than the Indians of Atabapo, or the Lower Orinoco: their plantations of the *zuca* shrub are very extensive, and the women make large quantities of manioco from the root. Indians of this tribe frequently visit the British settlements on the Demerara, taking advantage of the proximity of the head streams of the Ventuare, Caura, and Caroni. Many of the Marquiritare, who stopped to see me as they passed, pronounced a few English words very distinctly. They bring English trade-guns with them from Demerara, for the Spanish creoles, who purchase them in preference to the trumpery cocopetas sold at the German stores at Ciudad Bolivar. The Marquiritare are also one of the famous tribes for the manufacture of the urari poison, and the beauty and quality of their blow-tubes. They preserve the plumage of beautiful birds for their feather-work, hammock fringes, &c., especially that of the gallo de piedra (the cock of the rock), so conspicuous from the fine orange tints of the small birds, only seen in this district among the boulder masses forming the raudales of the upper courses of the rivers, particularly the cataracts of Maypures,

to face page 102.

H Wickham

ORINOCO BELOW THE MOUTH OF THE CONUCONUMO

and on the rivers Conuconumo and Padamo. They positively affirm that salt is an antidote to the poison, and say that if the mouth of an animal be filled with salt as soon as struck, it recovers, and speedily becomes very tame.

A young Spanish creole named Roja, with his two women, worked for me during February. I calculated a hundred trees for one man's tapping as the amount of his daily labour. A large herd of barquiro (wild hog) wandered about my water-enclosed piece of land. Sometimes they mischievously broke up the palm-leaf cups in which I caught the ciringa milk, and we occasionally secured a dinner from their ranks. With the addition of Roja and his women to my company, the roof of my rancho afforded scanty accommodation, although they always slept outside except the nights were rainy. Lately I had had the first touch of fever since leaving San Fernando; and about the 8th of February I began to suffer much from extreme nausea and vomiting, which preliminary attack came on in the forest, whilst going my round of tapping the ciringa. I was a long way from the puerto of the path where the canoe was secured, and had great difficulty in getting there, as each time the fit of nausea returned, I became quite powerless, and had to drop down on the damp earth, and wait until the paroxysm was over. When I staggered to my feet, my machete would get betwixt my legs, and nearly capsize me again. Having at length reached the curiara, I

endeavoured to paddle up the little branch creek
to my lodge; but the sun was too powerful for me,
and I had to scramble on shore again before I could
make the attempt to reach it. Fortunately, I was
now not far from it, as I was reduced to crawling
on my hands and knees, and the remainder of my
strength fast failing. However, eventually, I did
reach the bench made of split stems of the manac
palm I used for a bed. I remember little of what
passed during the four days that the constant nausea
and vomiting lasted. It is singular what an impression the slightest mark of kindness and human
sympathy makes on one in such an extremity. I
recollect one afternoon, as I lay prostrate and
incapable of moving, and part of my back bared
to the swarms of sand-flies which filled the air;
at that time a woman of Roja's entered, and
seeing my condition, she passed her cool soft hands
gently over my burning brow and back, brushing
away the plagues. Although unable to thank her,
I think I never felt so grateful for anything. The
Indians firmly believed my sudden seizure to have
been caused by a sight of "the little pale man of
the forest," whom they say is a little elfin sprite,
appearing occasionally to people alone in the forest,
rising from its abode among the roots of certain
trees which it particularly affects. When visible, it
is supposed to be the sure precursor of evil to the
unlucky beholder, if not of his death. They all
considered me at that time to be a doomed man.
As I was unable to eat anything procurable here,

my weakness increased. The want of breeze was also another drawback, for the currents of air that, from time to time, sweep down the broad Orinoco, do not reach the transverse bed of the tributary streams. Roja and the two women continued to tap the trees, bringing in a little rubber daily. As I became weaker, I felt that the only chance for me, and even that a small one, was to go and spend a few days up the river, on the more breezy shore of the main Orinoco, at the rancho of my neighbour, Merced Gil. He and his family were most kindly attentive, and I did get better. Strangely enough, the first thing that stopped the continuous sickness was a draught of *gaurapo*, made with the heated juice of sugar-cane. My host attributed my illness to my having drunk two kinds of water in the creek, Agua Negra and Agua Blanca. At parting he gave me some of his small store of the fine tobacco of the Cassiquiare.

Rogers returned from Conuconumo in, apparently, a very weak state, and said he had been sick all the time he was away. He brought with him a little manioco and tobacco, and more was to follow. It is when recovering from illness here that one regrets the absence of any beverage but water, and the accompaniment of unpalatable solids in the shape of crude flesh or fish, to be eaten with the coarsest description of breadstuff. I was compelled to abandon my rancho up the creek, it became so infested with niguas; and had another put up at the mouth, where a slab of rock slopes

down into the water. In order to escape, in a
measure, from the torment of mosquitos, I had this
one constructed with the palm-thatch down to the
ground all round, leaving only a small hole (over
which I hung a blanket) for entrance. Here, in
the dark, I could enjoy a little rest in my chinchora,
when I came in tired from the forest.

Roja caught a sloth one morning in the act of
swimming across the creek. This was the first time
I ever tasted the flesh of this curious animal, and
although it was badly cooked, it was really good
eating. Next day several fine wild hogs were shot,
but we had great difficulty in jerking the meat
during the rainy weather, for want of sun.

CHAPTER VII.

FEBRUARY 27TH.—The rains continued to increase in violence, and the river had risen greatly, notwithstanding that this was the dry season. For many days I was unable to tap the india-rubber trees, and Ramon was laid up with what is called "a game leg," and most of the other people were suffering more or less from calentura; consequently, I took very little ciringa.

MARCH 1ST.—Heavy rains were incessant, chiefly at night: the Orinoco was very much swollen. Merced Gil was swamped out of his work at Caricia, his ciringa trees and rancho being under water. This week we killed three of the larger kind of wild hog called *barquiro;* they appeared to me identical with the javiti of Central America: an immense herd of them wandered about the exterior of the rancho, and Merced came down to join in the shooting. After we had secured several, we stowed them away in the canoe. Benacio and the boy Narciso did not appear with the one entrusted to them, though we could hear them whistling at no great distance, and called to them repeatedly. I

suppose they were over-elated at the prospect of
their favourite meat for a feast; for though I summoned them several times, still they loitered.
Merced Gil was sitting in the curiara, and the sand-
flies were in clouds: I could stand it no longer; so
vacating the stern of the canoe, I jumped on shore,
and advanced along the path to meet the truants.
I suppose I did not look amiable, for no sooner did
Benacio see me, than he dropped the end of the
pole on which they were carrying the pig, and
bolted into the bush. In the evening, as he did
not return, I considered he had absconded altogether, although, from intimations I received from
time to time from Merced's wife, (who was staying
at my place during her husband's absence at the
Conuconumo,) herself an Indian of the pueblo Maroa,
I was certain that he was hanging about the place,
and was receiving food from the others. I never succeeded in catching him, though several times I rose
in the night and went by a circuitous route to the
men's quarters; but he was always too quick for me.
Ramon admitted he had been there, and was living
somewhere in the forest, and that one of the women
was generally with him at night. He afterwards
induced away the stupid boy Narciso, and I saw
no more of them. Roja completed his month of
service, and left for Maypures, which defection
nearly deprived me of hands, as Ramon was sick,
and able to do very little, and Mateo was with
Merced Gil. The peons of this district are hard to
procure as workers. They are almost all deeply in

debt to the principal creoles of the pueblo, and when they are secured, they are fit for little, as they have all the vices of a reduced and selfish race.

About this time I was greatly surprised at finding a French gentleman in my lodge when I returned from the forest. In so remote a situation all Europeans appear as countrymen. This traveller was journeying to Esmeralda, intent on exploring the country behind the great peak of Duida, having determined in his mind that there must be a mine of gold in that direction. Thinking to return this way in a month, he promised to call on me, that we might have a raid on the *barquiro* together, but I never saw him again.

As soon as Ramon was on his legs, we tapped the trees, with a little better yield. The water, after having risen to within a few feet of the door of my rancho, subsided as rapidly, and we had dry weather for a short time, just as we had begun to despair of it.

19TH.—I was again troubled with much fever at mid-day, but the attack was not sufficiently severe to prevent my getting through the tapping of my trees.

26TH.—During the past week the weather had been very fine ; but, owing no doubt to the stagnant water-pools, the forest now swarmed with the zancudos mosquitos, and whilst at work we literally led a life of torment. These zancudos were of a reddish colour, unfamiliar to me, and they bit dreadfully in the shade of the woods during the daytime, and came out in full vigour on moonlight nights.

I now sent Rogers down to San Fernando to seek advice of the padre of the pueblo, who enjoyed a local celebrity for physic. He went with Merced Gil. He had not been able to do one day's work in the forest for some time, and was in a very weak condition.

APRIL 3RD.—This was the third week of fine weather; but I found the position I occupied would not be tenable much longer, as I had no more ammunition for my gun, and had, therefore, to rely entirely upon the fish-hook. In the night Ramon sometimes took a small species of cayman, called *bavia*. I did not dislike the flesh. It is best salted, but it had an unpleasant odour and taste of musk about it,—resembling the flesh of some large fish more than that of an animal.

When it was fine I used to sleep on a rock on the bed of the Orinoco below, in order to avoid the zancudos. We were only once fairly caught by the rain, although we often had to take warning from the lowering sky and distant thunder, to get within timely shelter. The *playas* or sand-banks scarcely appeared this year; that opposite the mouth of the creek was only dry for two or three nights.

10TH.—The weather continued fine, and would have been really enjoyable, had it not been for the insect population. The nights were, for a short time, unexpectedly clear and bright for this humid region, and I was able to sleep well on the rocks below. During the brief paddle there after supper, I was often struck by the extreme grace and beauty of groups and clumps of tall macanilla palms, especially if the crescent moon were rising behind their leafy crowns. One morning, when sleeping on these rocks, I awoke as usual in the grey of the dawn, and saw a funny-looking little quick-eared creature perched on the top of a boulder. I could not make out what it was in the faint light, or how it could have got there, surrounded as we were by the river, but I secured it by casting my blanket over it. My prize proved to be a beautiful little animal, called by the natives "ravo pilado," on account of its tail being without hair. The rest of the body was covered with beautiful woolly fur. The one I caught had two imperfectly-formed young firmly attached, one on each side, under the fore legs, externally, after the manner of this curious creature. Its ears were transparent and delicately veined, and it twitched them, as if hurt, by any sudden noise. From the appearance of its large eyes it was nocturnal in habit, and not accustomed to be surprised by the broad daylight.

15TH.—GOOD FRIDAY.—Ramon had been unable to work again for some time past. Last Good Friday I spent in an Arigua village on the Upper Caura. This Lent I had no need to observe the fast, for it was of necessity: there was nothing but a little, very little, fish to be had, my ammunition being long expended.

In the afternoon, after tapping the trees, I used

to set Ramon and Manuel to work with hook and line; in the mean time, I paced up and down upon the dry slab of rock at the water's edge, in front of my rancho. It may be imagined that the line was watched with sufficient interest, as thereon depended supper and breakfast for the morrow. I did not lose the best hours of the morning, as they were given to the tapping process. We caught some very large tembladors (electric eels) in the pools of standing water in the forest. We used to spear them with long lances of sharpened saplings, as they lay concealed under the rotten logs which darkened the water. These pools also contained small fish of curious shapes. I was delighted at discovering that the sand-flies, those inveterate plagues of man in these regions, are not without their own enemies. My attention was at first attracted to a small fly thickly settling on the blanket that was suspended over the entrance to my dark rancho; and when I watched them more closely, I observed that each held a sand-fly spitted on its proboscis, which it had evidently secured on the wing from amongst the dancing myriads before the door, returning to the blanket to consume the captives at leisure. A diminutive but active yellow wasp also disported itself on the surface of the blanket, pouncing upon any of the sand-flies that became momentarily entangled in the hairs, speedily devouring them. It is a misfortune that these exterminators are not more proportionate to their prey! Most of the native fishing-lines, and the best, are those made from the fibre of the young, still-folded leaf of a palm called cumare. Other palms, such as manriche, milite, macanilla, &c., make good cord, but do not equal the cumare for strength and the endurance of water. Our strongest water cord is not to be compared to it. The finest chinchoras (or hammocks) are also made from this palm, though the other three varieties supply material for an inferior article; but these fine chinchoras are the "grass hammocks" mentioned by the coast travellers.

The rock at my rancho was a favourite resort for numbers of butterflies of different species, as all the rocks were that contained little puddles, alternately covered and uncovered by the rise and fall of the river. They settle in closely packed clusters of colour, and, when disturbed, mount cloud-like into the air, but soon re-settle on the margin of the pool.

17TH.—EASTER SUNDAY.—The long course of Caribee fish was at last broken. We took a *caharo*, a large fish, with an immense head: the flesh is substantial, and makes good salt provision. I very soon tired of the tembladors, though they were not bad, but of too gelatinous and viscous a consistency when cooked to be eaten constantly. I was taking a siesta in my chinchora, after coming in from the forest, when I was roused by a voice calling out that the curiara was adrift. True enough, I found the little rascal Manuel had not made her fast properly, and, the river having risen suddenly, she was going off in the current down stream, with the paddles.

I

It was well that I got out in time, as Ramon was not able to take to the water in pursuit of her, and Manuel was afraid of caribees. I must confess that after a sharp spurt I was not sorry to get my arm over her side and scramble in. On another occasion. (during a previous rise of the river), when we came in from the woods, I found the other canoe, the casco, gone. As I thought it could not have got far, I started off at once; but, though paddling hard, we did not come up with it until sunset, when we found it caught in some bushes at a bend in the river. Although very hungry, there was no alternative but to paddle back as soon as possible. We did so towards morning, through a deluge of rain, and quite famished. The rain now seemed fairly to have set in; the river, after having fallen somewhat lower than before, rose rapidly to within a few feet of the rancho door. Many different kinds of ranos (tree frogs) and ground toads (zapos) croaked loudly from the shore in as many different voices. The forest atmosphere was heavy with the fragrance of orchids, and other plants of the same nature, unfolding their flowers to the increasing moisture that hung in the branches of the trees. Owing to the turbid current of the rising river, fish became very difficult to procure, and the rains rendered it impossible to work in the forest with success; lagoons of standing water crossed all the paths. At the end of the month I evacuated my position as no longer tenable, intending to make a push for Pará by way of the Cassiquiare, Don Carlos, and Manáos.

26TH.—Two days were passed in making a palm-thatched toldo for the casco (the ciringa trees had been tapped for the last time), and we freighted her with india-rubber, &c. I had, with the assistance of the little rascal Manuel, now to do everything myself, as Ramon was quite off his legs. The rancho was entirely swamped, the rains becoming so heavy that the water draining off the forest-covered slope behind ran through in a continuous stream. It was impossible to get out of my chinchora at night without splashing into the water above the ankles. Snakes and zapos began to be unpleasantly numerous in the thatch. Many of my rubber trees were under water in the forest, and when it was dry enough to tap the remainder I could not reach them without wading waist-deep. It had often puzzled me where the little fish to be seen in this freshly fallen rain came from, as the pools and lagoons had no connexion at all with the river.

27TH.—Unable to start early. It had rained in torrents all night, continuing on into the day. We had had nothing to eat but manioco for three days past, and little Manuel looked quite pale when we started. Ramon was helplessly sick.

28TH.—After a hard struggle up stream against the current, I found it was out of the question to make sufficient headway with the two heavily loaded canoes, and little Manuel only to assist me. I therefore made the casco fast in a sheltered place, and pushed on in the smaller curiara to solicit aid from any ciringoros I might chance to find still at

work on the river above.   After a long day's paddle
against the stream we found the rancho (the only
one Ramon knew of) had been abandoned.   The
trees of this ciringal had evidently been very carefully tapped, from what I saw of those near the
rancho.   I borrowed a wrinkle from these people,
which I may adopt with advantage at some future
time.   With great reluctance I gave up the idea of
passing by way of the Cassiquiare, now but a shoit
distance from us, and instead, gave the word for
down stream; but it was impossible to conquer the
current with the two canoes and no fit coadjutors,
although if we had been able to gain the Orinoco
mouth of the Cassiquiare, all our labour would have
been over, passing into its swift current.   No time
was now lost in cooking, for we had nothing to
cook!   We soon reached the casco again, and then
on, past Caño Caricia.   I took a long look at the
forest about its mouth, which had become so familiar
to me.

We found some of Señor Lanches' peons still at
Purunaminare island, but they were about to leave.
I was very glad to get a piece of gritty salt fish
from them.   We passed Serra Yapacani in the
evening.   It was very calm, and the sunset was
beautiful.   The surface of the river seemed to resound with the noise of a fish, the *bocon* of the
natives.   We startled immense flights of gauzy-
winged insects, passing through them as we went:
they were flying obliquely across the river with
the greatest regularity, all in the same direction.

In the middle of the third day, after having turned down stream, I arrived at San Fernando de Atabapo. After getting clear of the raudales of Santa Barbara, I had lashed the two canoes together that I might continue on with the current during the night whilst we slept, as there were no rocks or rapids to fear below. I was very glad when we got safely below Santa Barbara, as the casco containing a great part of my rubber was too much for little Manuel to manage among the rocks and runs above.

Although for many days together I had taken nothing but "incuta" (manioco and water mixed), I think I never felt better in my life. In spite of alternate rain and a fierce sun, I did not suffer from thirst.

At San Fernando I found a change in the government, Señor Lanches, the creole white from Barquisimeto, replacing Castro as Governador. Rogers was at the village, looking much better. I took up my lodging at the hospitable house of Señor Angel Maria Oveidos. He was going into Brazil, and we arranged to journey in company. During the delay I again enjoyed the abundance of the delicious oranges of San Fernando, and I saw some of the beautiful rich red pigment called *chica*, which is prepared chiefly by the Poyñaves of the Mirida.

I think I have never experienced so relaxing a climate as that of San Fernando. The water of the Atabapo has a glassy smoothness, rarely ruffled by a breeze. A small glimpse of the river, seen from the door, reminded me of the Thames above Richmond on a sultry summer's day : long islands of low brushwood divide the placid stream. I had at this time a severe attack of the nausea and vomiting, which speedily reduced me to a very weak state again. I therefore acted upon the advice of my host, Señor Angel Maria, to start on before and await him in more salubrious air at Javita, or elsewhere.

The first pueblo after leaving San Fernando is called Chimuchin; afterwards, those of Baltazar and Santa Cruz appear; and finally Javita, situate on one of the head streams of the Atabapo, on the low division betwixt the waters of that river and those of the Rio Negro. Here everything had to be carried about nine miles overland to the stream Pimichin. The villages of the Atabapo and those of the Guarinia above the Cassiquiare, supply nearly all the peons employed by the creoles of the country. The Atabapo district is, indeed, a land of water; at the rainy season scarcely any dry land is to be seen, except that occupied by the Indian pueblos. The saying of the Indians, that "where the waters are black the stones are white," is fully verified by the bleached look of the rock on the shores of the Atabapo and its equivalent, that where the waters are white the stones are black— equally so by the polished surfaces so characteristic of the Orinoco. The utter absence of water-fowl denoted the scarcity of fish, and the traveller on the Atabapo has the greatest difficulty in getting

sufficient provender for sustenance. After passing
Santa Cruz, the canoe-men constantly left the main
stream, and followed well-cut water-paths, just wide
enough for the canoe to pass through the flooded
woods. The trees were of little variety, and mean
growth. For the most part they are identical with
the kind producing the exceedingly light wood,
found bordering the black water lagoons, through
which flows Caño Caricia. There are also those
clumps of the little palm with fruit and leaf resembling the manriche; each stem bears a crown of
five or six leaves. On the shores of the Atabapo
and its head streams, the big timber is seen only in
the distance. Amongst them, and deeply flooded,
I noticed the lofty tree with its dome of violet
flowers, that I had observed on the Upper Caura.
The chiquichiqui palms abound. They supply the
piassava of commerce. At length we drew up at the
landing-place of Javita. This pueblo is particularly
conspicuous, from the great number of pijijan palms
growing in clumps about the houses. San Antonia
de Javita is named after the last native chief of the
Guainia. Afterwards, the Indian tribes of that
river (except at its sources) and of the Atabapo
were amalgamated, and blended together in pueblos
established by the old Spanish missionary padres,
who penetrated into the country on the retreat of
the Portuguese traders lower down the river, and
the advancement of the Spanish boundaries. They
have, since the *exeunt omnes* of the padres, become
veritable peons, and appear to have lost all know-

ledge of original nationality. The Baniva tongue is pleasant to the ear, and is spoken in all the pueblos from San Fernando de Atabapo to San Carlos de Rio Negro. The sand in the vicinity of Javita, and indeed of the whole Atabapo district, is of such snowy whiteness, as to form a great feature in the landscape. This is especially the case in the dry season, when one is able to walk for miles along the river's course on broad, hard sands, of dazzling whiteness. The sandy ground (which is very wet) affords birth to a great variety of ferns and mossy plants, and others with lily-like leaves, which combined, give a peculiar character to the vegetation. One very closely resembled our common English brake fern, but it was smaller. The sight of it raised up in my mind an affectionate remembrance of home.

I waited some days at Javita, but as it became increasingly difficult to obtain anything to eat, I determined to push on a stage to Maroa, a village on the Guainia, or main Rio Negro. The portage path through the forest was well cut, which was an agreeable novelty, to be able to walk comfortably, and at the same time to be able to admire the beauty around. For the greater part of the way the path was hard and dry, a succession of slight rises and falls. In the latter we crossed many little streams of clear water, but stained with the inevitable coffee colour so characteristic of the district. These little streams run in different directions; that is to say, the water of some ultimately finds its way into the ocean from the Amazon, those on the other hand by the Orinoco. They are crossed by long logs, nicely squared with an adze by the Indians of the pueblo, whose duty it is to keep the portage path in order. These squared logs are also laid down wherever the soil becomes boggy, so as to afford firm footing to the Indians engaged in the transit as porters. Altogether it is quite a creditable affair—and the nearest approach to a road I have seen here. I passed the night at the puerto of the path on the Pimichin. Next morning I woke at the first crow of the cocks from the solitary lodge. I never now hear. the crow of a cock ringing shrilly out on the morning air, but I think of many a start I have taken in some dim, unknown, moonlit path, in the Mañana por le Madenga before dawn. I verily believe it is *one* of the impressions I shall never lose. I suppose we all have some impressions of this palpable, and at the same time, vague, and undefined nature. The Pimichin, must be I imagine, the most winding and circuitous of all streams, and though narrow, it is very deep and rapid. We embarked in a very crazy old canoe, and on our passage down stopped to bathe from a rock. Making a sudden plunge out of the sunshine, I was quite startled at the extreme coldness of the water (it is proverbial among the natives) ; this quality seems scarcely to be accounted for by the stream having its course through the shade of a sombre forest, and beneath an almost always clouded sky. We arrived at the pueblo of Maroa on the 24th of May. Señor Andreas Level received me courteously, and prepared a banquet of such viands as were procurable in the place. I had been fortunate in having had two such very fine days, rare in this land of rain, and had I been stronger I should much have enjoyed the walk across the portage.

CHAPTER VIII.

I LITTLE thought, on reaching Maroa, how long I should be delayed within its precincts. The whole of June was consumed in the completion and the loading of the lancha of Señor Angel Maria Oviedos and Andreas Level with chiquichiqui (piassava), sarsaparilla, and the rubber produce of their ciringals.

I endeavoured to pass the time as best I might, by visiting the pueblos below San Miguel and Tomo- Terikin, but found them alike without interest; in fact, there appeared to be but one source of enjoyment here, that of bathing in the delightfully cool and limpid water of the Black River, very different to the turbid Atabapo, which seemed rather to relax the frame than to invigorate and refresh it. Passing from the Orinoco to this river the change of climate is very perceptible : the natives complain of cold; indeed, it is often chilly, especially at night or in the mists of the morning, when one finds comfort in gathering a blanket about one in the chinchora. I caught a cold at Maroa, for the first time during my tropical travels, owing, more probably, to greater susceptibility, caused by my late insufficiency of nutritious food, than to the alteration of the temperature.

The dark waters of this river are generally over-hung by a cloudy sky, and the reflection of it throws a deep shade upon the green of the foliage, giving a sombre character to the land, very different to the variety of colouring that is displayed on the forest bordering of the neighbouring Orinoco. Nor is this the sole difference. The forest of the Black Water contains little fish, and harbours less game. At the same time, the plague of insects, ever present on the White Waters, is not a drawback here. In the Rio Negro villages the chief theme of conversa-tion on meeting is the "hambre" of what was last eaten, together with speculations as to what will turn up for the next meal. Even the chief creoles of the pueblo, that is to say, the people who have the most command of the peons, are all more or less in the same strait. They have only them-selves to thank for such a state of things : they keep these peons, who are all most deeply in debt to them, constantly at work, chiquichiqui (piassava) cutting, rubber-collecting, or boat-building, till there is no time left for anything else in the way of such industry as planting or stock-raising, which is never thought of. They rely for manioco chiefly on the neighbouring Indian tribes of the Mar-quiritare and Poyñaves. Almost without exception, these people are refugees from the more populous districts of the republic. Each of the principal pueblos, San Fernando de Atabapo, Maroa, and

San Carlos, possesses a sort of residential clique, having almost a monopoly of the surrounding districts. At San Fernando, for instance, Señor Lanches, present Governador, Castro, ex-Governador, Andreas Level, the padre, and my old friend, Angel Maria Oviedos, have the chief command of the Indian labour, and consequently the greatest ability to draw the natural riches from the environing forests; at Maroa, Andreas Level and the Buenos, with the stout Don Juan at their head; at San Carlos, the Calderons.

I formed a friendship with Level during our descent into Brazil, which circumstances afterwards only tended to cement. Coming from one of the best Venezuelan families, he had been educated in the turbulent spirit and traditions of his country. After serving, and being often wounded in the endless revolutions, he had thrown up his commission, and had bent all his energy (he was singularly energetic) on opening commercial relations with the coast. He was fond of relating to me how, a few years ago, he and his brother Pedro had started up the Orinoco alone, in a curiara, with nothing but their shirts and guns; how he had succeeded in opening trade relations with merchants of Angostura and Pará, and cleared up a fine ciringal on the Cassiquiare, which he and all his peons worked in the dry season. The remainder of the year was spent in making lanchas at Maroa, and in collecting piassava, chinchoras, &c., for the next journey to Pará or Angostura. He had recently

brought the members of his family together in
Maroa: his father was an officer of rank, and a
Venezuelano of quite the old school; he had fought
with Paez, and had grown old in the service, but
the beggarly republic remained in his debt. In
order not to be entirely dependent upon Andreas,
who was the prop of the family, the old gentleman
worked at tailoring with a Yankee sewing-machine
he had brought with him; and throughout the day
the industrious whirr of the machine might be heard
in the quiet pueblo.

Andreas Level assured me that he had seen a
species of dog, *perritos*, hunting in packs about the
district of Maguaca. They were smaller than the
ordinary dog of the country, and had very pretty
long light brown hair and tail. The Indians of the
Maguaca sometimes caught them when young, and
tamed them.

I found the Venezuelans of the superior class a
singular mixture of qualities, pleasing and repulsive.
There is amongst these people what we should call
an absolute disregard of common decency: they
seem to be utterly ignorant of its very existence.
When I first became acquainted with Level, he
treated me somewhat distantly—inviting me to his
table only when he could place a fair spread before
me, and then, sitting by, would scarce touch anything himself. When he began to entertain feelings
of friendship for me, on the other hand, it seemed as
if he could not be too disinterestedly generous in
his attentions. At Maroa I received many kind-

nesses from a certain Señora Delfina. She gave me many little entertaining anecdotes of a tall Englishman, "Un Englez muy alto." How, when she was still a girl, and when she was living with her family lower down on the Negro, he had visited their neighbourhood, collecting and carefully storing away butterflies and all sorts of curious creatures. How several times he had, by making her look through a strange little instrument, converted little harmless mites of insects into terrible monsters. She seemed to think it odd that I did not know him personally. These people fancy all Englishmen are known to each other as well as they are among themselves. I concluded, however, that Doña Delfina's Englishman might have been the naturalist and traveller Wallace, who ascended the Rio Negro and its little-known tributary, Uripes.

The Indians of the pueblos, nominally Christians, still adhere to many of their ancient customs: the festas held professedly in honour of the saints are mixed up with their old dances and pantomimes. A very curious one occurred whilst I was at San Miguel in the month of July. The Indians of Maroa had been invited, and came to it in a fleet of canoes, and the whole population of San Miguel turned out to meet them on the stony *plaza* in front of the mud church, where they speedily formed into a long string of dancers, men and women alternately, and then wound like a chain around an offering placed on the ground by some girls. Each man had his chinchora slung across his shoulders, the woman behind him clinging to it with both hands: the two foremost men had long trumpets, decorated with feathers and beads, which produced long, low notes, constantly directed towards the ground in the centre, as if to exorcise some spirit or genii. The others who followed were provided with an instrument like a mallet, only hollowed out inside; these, as they were struck regularly on the earth, gave out a measured sound resembling a drum, and accompanied the pace of the whole whirling circle in their rhythmical chant: the language used in these chants is understood only by a few of the initiated. Every time the chant seemed to have reached a climax, it was wound up by a wild cry from the women. It was surprising with what pertinacity they continued to keep up the heavy measure of foot and voice: from its peculiar wildness and mystery, this dance would be invaluable if adopted for the witch scenes on our stage.

At Maroa the festival of San Juan was celebrated in the evening by the lighting of bonfires around the square of the pueblo. All the younger Indian men and boys formed a line and careered madly in a circle, leaping the fires in succession; those most inebriated simply stumbling through, and scattering the burning sticks and sparks far and wide with their bare feet.

The natives compound a variety of refreshing beverages by mixing the grated fruits of palms with water, and if it be an object to appease hunger, as is generally the case on the Negro, incuta (manioco and water) is added. The most universal drink is that made from the fruit of the manac, the assai *(Euterpe oleracea)* of Brazil, the seje, and the chiquichiqui, or piassava palm. The two latter give a very milky look and consistency to the water. The piassava has a beard-like growth, which drapes over, hiding the entire stem of the palm; as the tree increases in height, it appears at the base of the leaves : it grows in community, but only in the Aguas Negras districts. The natives affirm that the growth of this palm is exceedingly slow, hardly perceptible in a man's lifetime; but the stone of the fruit readily germinates in the ground, and they are everywhere to be seen freshly rooted about the houses in the Indian pueblos. Palms differ greatly in matter of growth. The natives say that the *pijijan* quickest reaches maturity; the manac is also quick; but the manriche is a very slow-growing tree. The large-headed turtle, called by the Indians *cabason*, seems peculiar to the Aguas Negras : its jaws are very powerful, and when captured it bites furiously at anything in its way. It is considered the best eating of all the South American turtles.

After a protracted delay in loading and drying chiquichiqui at Tirikin, we passed down stream on the 2nd of August. The peons of Tirikin are inclined to be more turbulent and less easily managed than their fellows in most of the pueblos. The traders appear often to have some trouble in getting them to bring in the piassava (for which they have

K

received goods in advance) without a liberal expenditure of rum.

The chief, or only industry of these villages,
especially of Maroa, is the building of lanchas,
to be sold at Angostura or in Brazil. Many of
them, however, come to grief in the passage of the
raudales. It is really astonishing to see the
Indian carpenters finish one of these large, well-
made lanchas out of the splendid wood of the
country (paraturi) from first to last, with little
more than axe, adze, and compass.

At San Carlos we were hospitably entertained
at the house of Señor Calderon. The village, with
the exception of the dwelling of the above-named
gentleman, had a mouldering look; it was surrounded, like most of the other Rio Negro villages,
with thickets of coffee, growing wild. The plaza
was dotted with groups of sheep. There were also
a few fine heads of cattle to be seen ; they belonged
to Señor Calderon, who appreciated the luxury of
milk with his coffee—a rare thing on the Rio Negro.
The sheep were poor. After passing the *remolino*
at the junction of the Cassiquiare, the foliage on
the banks became much more varied, and I recognized the now well-known leaf of the ciringa. We
formed quite a flotilla—two lanchas of Señor
Oviedos' and three of Level's. The larger of
Level's, the "Guainia," was the most considerable
craft that had passed down the rapids. At Cucay,
the frontier post of Brazil, there is a stockade and
a few *papuya* soldiers, commanded by a gentlemanly

man, dressed in the sensible light brown linen of
the Brazilian army. Here our passports were
looked at. I may mention that this was the first
and last time mine was demanded throughout my
sojourn in Brazil. I went to the office and gave
it up before I left Pará.

The Rio Negro becomes very broad, and is
divided by innumerable islands of all sizes, and
the lofty mountain rock, Cucay, raises its crests,
in the ledges of which a suspended forest, as it
were, appears to hang above the mists far away
on the left bank.

The natives seem to have a decided predilection
for making their *sitios*, or small plantations, on the
islands: the number of these sitios greatly increased
after our passage within the Brazil frontier. Leaving
the mouth of the chief tributary of the Upper Rio
Negro, the Uanpes, on the right bank, we soon came
within sound of the chief rapids—those of San
Gabriel. The scenery of this neighbourhood is
beautiful. We passed down with nothing worse
than a broken rudder: such good luck, however, is
not always the case. One of Señor Calderon's
lanchas, that had been despatched a day or two
before us, was totally wrecked, and the bales of
piassava lay strewn on the sands below. I descended the cataracts in the lancha of Angel Maria;
he and Level following in the others. As the rudder
unshipped, she went broadside into the heavy rollers
at the foot of the chief *cachuera*. Fortunately, we
floated down with only a complete ducking! When

we had fairly brought up in a little cove, and secured the lancha to some bushes, I embarked in the canoe with the Indian patron (the same man who had brought Castro's lancha up the Orinoco), and paddled round the foot of the cachuera to the village of San Gabriel. The short, heavy, spoon-shaped paddle used in this river is very awkward compared to that of the Caribs, to which I had become accustomed. Indeed, I was really tired by the spurt across the eddying water, although I was in good condition for such work.

The view from the crumbling parapet of the old fort on the heights of San Gabriel is particularly imposing, backed north and south by high hills of varied outlines. The village itself is very miserable in appearance, though it is the post of a comman-dante and some soldiers. After enjoying the prospect of the rapids for some time from the old ruinous portaleza, I was wandering among the straggling, tumble-down huts of the village, when the com-mandante called me into his house and insisted on my taking dinner with him, which I thought was very hospitable, for in my costume of shirt and trousers, and they none of the newest, I must have looked rather disreputable. The following day, whilst awaiting the descent of the lanchas, I took a bathe in the river above the rapids. The bank seemed here to slope abruptly to a great depth, and, although not a feeble swimmer, I took care not to let go my hold on the tough herbage, as I had bought experience of the strength of the floods that look so

smooth before they glide on to the rocks of the
cataracts ; with this precaution, it was very pleasant
to feel the water seething round the limbs in violent
ebullition: one might almost imagine the sensations
of a fly involved in a tumbler of soda-water ! The
waters of the Rio Negro in its upper courses are
always delightfully cool. At San Gabriel we enjoyed
the unexpected treat of fresh beef. Some forty head
of cattle and a few sheep graze on the heights.

Below, the river continued to increase in breadth,
but the water became somewhat charged with
floating particles, less remarkable for purity than
those of the Guainia or Upper Negro.   On the 19th
the river became blocked by islands and islets
forming the runs of Massarambé.   Beautiful scenery.
Sitios very numerous in comparison with the
Venezuelan rivers.   I felt impressed with the con-
viction that when Brazil, the Empire of the South,
shall have developed, and grown in her Amazonian
provinces, she can easily absorb the Rio Negro
districts to the great cataracts of the Orinoco. What
a future must be in store for Brazil, if her Government is guided by a firm and judicious hand !

Our Indians now remarked that the woods
assumed much the appearance of the Orinoco, without the same animal life—the aregnato, howling
monkeys, game-birds, &c.   Fish was difficult to
take. Below Massarambé the river again opens
out into a wide expanse, but at the little village of
Castanhiro is divided into channels at the bend,
formed by the hills of Gruya.   These hills resemble

the eminences which characterize the Upper Orinoco; but it is doubtful whether to designate them as mountains or as immense rocks, they are so like huge single stones in shape.

On the morning of the 20th the " Guainia" went aground on a *playa* above the little village of Santa Isabel, and it was with two days of much labour and the landing of much of the cargo in the forest that we got off again. We saw a great deal of ciringa in the forest, but the trees had been much tapped, and did not appear very productive.

The tree, the bark of which produces the *tabaré*, so much esteemed by the Indians of the Rio Negro for cigaretas, was here very abundant, and the bark of very fine quality. To prepare it for use, it is first cut from the tree in long narrow strips; while still green it is held upright, and the end is beaten sharply with a stick till it falls over in thin papery flakes; it is then carefully dried in the sun, and cut into convenient lengths. From its excellence, I should think it is likely one day to become an article of commerce. The Indians seemed particularly partial to the fruit of the alfaroba tree, the edible part of which consists of a sweet flour-like powder covering the very hard beans, that are enclosed in a large woody pod.

On the 24th there was an horizon of sky and water; the right bank became high, forming a clay bluff or barranca. The river was still deep, but in the dry season there are extensive sands here. The india-rubber of this river, unlike that of the Upper

Orinoco, is found only on the islands, and the few trees I had an opportunity of examining appeared to possess little milk in the bark. Level and I went ashore at the sitio of Chibaran (beautifully situated on a high bluff point), where we received an acceptable present of fresh beef. Below this we again traversed island after island, and were once again aground, but got off with less trouble than before. The river opened out to an immense width, and the banks being low, were unoccupied for some distance by sitios. The people say that one may traverse for three days (without reaching the mainland)—a maze of water, islands, and lagoons. The main channel appears generally to tend towards the right bank. In the dry season much of the river-bed is filled by extensive sands, to which the turtles resort to lay their eggs. The voices of birds became more frequent in the forest, and the take of fish increased. Twice we met animals crossing from shore to shore —a tapir, and a herd of wild hogs. We continued to coast the right shore, which again heightened into bluffs. At Moréra, sitios became more numerous. The reddish colour of the cliffs of Moréra formed a most pleasing contrast to the green foliage.

27TH.—This day we passed the village of Barcellos, once of importance on the Rio Negro. Like the rest of the poblaçons on this river, it has a very neglected appearance. The greater part of the young Indians of the Brazilian Rio Negro had been drafted away to serve in Paraguay, where the mortality amongst them had been excessive, a mere

fraction returing to the land of their birth. The
sitios have an air of much more ease than the
poblaçons; indeed, the natives bestow small attention
upon the villages they seldom visit, except for the
festas, but cultivate the plantations, where it is not
so difficult to obtain provisions.

On the 29th we passed the mouth of the largest
tributary of the river, the Branco, having its head
streams in proximity to those of the Orinoco tributaries, Caroni, Caura, and Ventuare; and also those
of the Essequibo. As at the junction of the Caroni
with the Orinoco, the white and black waters do
not commingle, but form distinct currents. On the
two opposing shores there is a strongly-marked
difference in the character of the woods. That
occupied by the black water has the snowy white
sands and the forest foliage of the Aguas Negras of
Venezuela, whilst the other shore reminds the
traveller of the Upper Orinoco. Immediately above,
the Rio Negro, in its breadth (more than fifteen
miles) and number of islands, is more like a lake
than a river. At the mouth of the Branco it is
comparatively narrow.

The lanchas floated on slowly down with the
current, and leaving the patron in charge, Level,
Angel Maria, and I often jumped into the canoe,
and paddled away to different sitios and poblaçons
on the shores. I well remember how, on the
evening after passing the mouth of the Branco, we
were caught in a *temporal*. We saw it distinctly
approaching up the great expanse, the water

whitening beneath the strength of the squall. We struck a short, quick stroke in order to reach the shelter of the big-timbered forest, but it neared too rapidly; so we put about into some thick, flooded bushes, close at hand. Here we allowed the brunt of the storm to go by, and then pushed on again, in the hope of finding the landing-place of the sitio. However, the darkness fell fast; and drenched, hungry, and tired as we had become, the prospect was not a cheerful one, although we endeavoured to keep up the semblance of hope to each other. Twice we fancied we neared the village. Level had been there once before; though at length we came to the conclusion, from the appearance of the water, that we were not following the main bank, but one of the interminable lagoons or back waters common to the river. On and on we went, and at length retraced our way; when nearly done up we descried a star-like speck, which we recognized as the light of the large lancha, and we eventually reached her, sopping wet, and utterly exhausted. After changing our clothes we supped in the moonlight, for the night was now fine, on the top of the toldo.

The highlands about the head of the Branco will probably be one of the richest and most beautiful districts of Equatorial America, when once fairly opened by enterprise. Lower down, immense herds of cattle wander over the country; they might be had almost for the catching, and if brought to Manáos (a short journey for the steamer), would immediately realize a large sum. Turn which way you will in this, the greatest water valley of the world, there seems to be an unlimited opening for enterprise or capital.

We went on shore at the dilapidated village of Airon, but found there only one old man and family, and nothing to be bought, all the inhabitants being away in the ciringal forest. In the afternoon the heat became very oppressive; it was often followed in the evening by heavy thunder squalls, in direct opposition to the cool temperature of the head stream.

We arrived at the town of Manoás (La Barra) on the 3rd of September. The town has a very thriving appearance; is situate on ground of a very unequal surface. Two of the large English-built Amazon steamers lay off the town, and looked very suggestive of a return to civilization. I shall now, therefore, cease these Notes of a Journey through the Wilderness. I will but say to those who contemplate bending their steps to the Tropical West, that after all my experience of Tropical America, I have come to the conclusion, that the valley of the Amazon is the great and best field for any of my countrymen who have energy and a spirit of enterprise as well as a desire for independence, and a home where there is at least breathing room, and every man is not compelled to tread on his neighbour's toes. I purpose to make the table-lands in the triangle betwixt the Tapajos and the Amazon, behind the town of Santarem, in future the base of my operations.

to face page 138.

OUR FIRST (Temporary) HOME, NEAR SANTAREM, 1871.

I parted from my Venezuelan friends, Level
and Angel Maria Oviedos, at Pará, and they returned to their distant homes on the Guainia. I
have since seen the following paragraph from a
Pará paper, from which I fear poor Level has come
to grief, on his way up the Negro with his goods,
and the results of the sale of his lancha and
produce:—

" Mais uma atrocidade dos indios.—No dia 9 do
corrente mez de dezembro, n'uma das margens do
Rio Negro, limites da freguezia de Moura, acabão os
indios Jaupery's (Uanpes) de assaltar uma canoa
do Venezuelano André Level que ha ponco seguio
desta capital para a republica de Venezuela, ferindo-
lhe gravemente quatro tripolantes e roubando-lhe
mais de cinco contos de réis de mercadorias, de
que ia carregada a mesma canoa. Pessoa autorisada
da mesma freguezia diz-nos que muito receia que os
indios vão á mesma freguezia e commettão os
maiores attentados, visto o abandono em que estão
de forças para repelil-os; e pede-nos a publicidade
deste facto para que providencie a autoridade."†

† *Another Atrocity of the Indians.*—On the 9th day of the
current month of December (1870), on one of the banks of the Rio
Negro, in the limits of the *freguezia* (parish of Moura), the Jaupery's (Uanpes) Indians attacked a canoe of the Venezuelan André
Level, who had a short time ago started from this capital for the
republic of Venezuela, seriously wounding four of his crew, and
robbing him of more than five *contos* of réis (£500) worth of goods,
with which his canoe was laden.

A trustworthy person of the same *freguezia* tells us that he

I was the more surprised on reading this, as the Indians of the river had always been considered remarkably peaceable. I suppose they must have become hostile from the system of kidnapping the young men for service in the Brazilian army.

much fears the Indians will return to it and make more strenuous attempts, seeing the utter destitution in which the inhabitants are of forces to repel them, and asks us to give publicity to this fact, that the authorities may be enabled to provide against it.

A JOURNEY

AMONG THE

WOOLWA OR SOUMOO

AND

MOSKITO INDIANS

OF

CENTRAL AMERICA.

143

PART II.

A JOURNEY among the WOOLWÁ or SOUMOO INDIANS of Central America.

CHAPTER I.

THE Bremen schooner "Johann," 350 tons, in which I was a passenger, sailed from London on the 5th of August, 1866, and after a tedious voyage we sighted the island of St. Lucia on the 4th of October. It appeared in the distance quite a pile of mountains, and we were soon able to distinguish the character of the scenery, for we passed very close to the S.W. point: the hills were of a most fantastic shape, and densely wooded to their very summits; a light breeze blew off the shore, bringing with it a delightful perfume, which, after my long confinement to the ship, made me desire a ramble amongst the woods and bright green slopes, that looked so attractive in the sunlight; and when, as we stood out to sea, the sun sank behind the Pitons, the view was indescribably beautiful.

Wrapped in a rug, the sleeping on deck was pleasant in the delightfully pure air, and the stars that thickly studded the sky looked like little lamps swinging overhead with every motion of the ship. We often sailed along through immense shoals of flying-fish, which, in full flight, strongly resembled dragon-flies. Many came on board, chiefly at night, flying over the sides of our heavily-laden vessel, when they were eagerly captured, and served up as a very welcome addition to our salt sea fare; indeed, they are exceedingly rich and delicate in flavour, more like smelt than any other fish. On the 21st we sighted the mainland about Monkey Point, and shortly after hills and islets covered with vegetation appeared in succession.

The approach to Grey Town is prepossessing: the hills being well wooded, and the crests of many of them are crowned with umbrella-shaped trees of great size. The mountains in the background as we drew near were partly veiled in mist and cloud. Shortly after sighting Monkey Point, a handsome butterfly flew on board, which I secured. It proved to be a *Heliconius galenthus.*

In approaching Grey Town, the captain, according to the sailing directions on his chart, very nearly ran on the bar; but, fortunately, a friend who had come to meet me, hailed him from a boat, and prevented the mishap that might have ensued. The mistake occurred through the constant shifting of the bar, arising from the quantity of sand brought down by the San Juan river; in fact, old Moskito men have told me that they can remember having seen English men-of-war lying where Grey Town now stands. The only vessels we found here were the American mail-steamer and an Italian brig.

I landed with the captain next morning, when he called on his merchant; and in the evening I brought my baggage on shore, and took a turn by moonlight among the houses, that were scattered in clumps. The riot of insect life was very remarkable: some sounds resembling the tinkling of bells, others, the chirrup of our own grasshopper, only much louder; frogs shrieked on one side and on the other made a noise like a kettle-drum.

One morning I took a delightful ramble on the skirts of the forest; the butterflies were very numerous and beautiful, varying from the size of a bat, to that of our very smallest species. The woods in all directions were traversed by the beaten roads of ants, called by the natives " weewes," along which the little creatures were marching with pieces of leaves in their mouths. I did not feel the heat as much as I had expected, it being very showery, as is usually the case at this season of the year.

The town is situated on very low ground, with a lagoon in its rear, and is altogether a very uninteresting place. The hills, visible from the sea, are lost sight of on landing, on account of the dense woods that intervene. The population is of an exceedingly mixed character, the Indian blood being

L

discernible in the warm coppery tinge of the complexions of some of the inhabitants, while others are as black as ebony. Many of the Moskito people speak very good English. It appeared strange to me, as yet unaccustomed to the ways of the country, to see the ladies promenading in the evening with handkerchiefs on their heads, and sometimes nothing at all, cigars in their mouths, spitting in the most approved fashion as they walked. On the evening of the day following my arrival, as I strolled out, I paused in astonishment beside a pool, listening to sounds created by the frogs; as for my English terrier, Jack, he was perfectly bewildered. I had purposed engaging a canoe that day, and starting for Blewfields, but waited till the next, as, by so doing, I secured a passage in a little Moravian schooner, "Messenger of Peace." I experienced much annoyance at having to pay duty on my powder, for, by some mistake, it had not been entered on the ship's papers as my own. I discovered also that English bank-notes were not available as current money on this coast, and was obliged to send those I had with me back to London.

On the 26th of October the "Messenger of Peace" started for Blewfields. As we passed the old "Johann," the captain waved his hat, and I was glad to see he did not harbour any remembrance of the little differences we had had during our outwards passage. We were rather a long time in reaching Blewfields, being detained a whole day about Monkey Point, under which we anchored, spending the time pleasantly enough, fishing at bottom for "crook-crook," so named by the Moskitos, from the noise the captive makes when taken out of the water: in appearance, it is something between the perch and the wiasse of our coast. We were becalmed the greater part of the next day, and caught more "crook-crook" for breakfast and dinner.

Monkey Point is certainly a very pretty spot, and would, no doubt, make a good harbour with Captain Pim's proposed improvements. At present, the Carib boats and other small coasters always make for it at the first threatenings of a storm, and are often for safety compelled to remain there for days.

About three o'clock, passing between Blewfields Bluff and Cassada Cay, we entered Blewfields Lagoon. The bluff is a bold headland, indicating the mouth of the northern entrance of the lagoon. From a distance it looks like an island, being joined to the mainland only by a low narrow strip of mangroves. After sailing half way across the lagoon, we anchored at sunset near a little island covered with broken cocoa-nut trees, called Bluffway Cay, and situated opposite the settlement. What struck me most here was the desolate appearance of the forest, ravaged by a hurricane, which had caused the devastation about a year before (1865). I went ashore at once in the canoe, intending to call on the missionary, Mr. Lündberg, to whom I had brought a letter of introduction. It was dark when I landed, and, as at Grey Town, I was much struck by the wonderful variety of sounds among the insect and reptile life. In coming on shore, my eyes, ears, and nose were quite filled with small things resembling sand-flies: these, however, did not bite, though they were annoying. I think they must have been peculiar to that time of year, for I did not see them afterwards in any numbers. Mr. Lündberg was away from home, having gone to Rama Cay when I called at the mission-house, but I was very kindly received by the gentleman in charge. Next morning, as I was standing on the little jetty, waiting for my baggage to be brought on shore, a slight little fellow, who was standing by, asked me my name. Having answered him satisfactorily, I inquired his in return, to which he replied, "William Henry Clarence." Shortly after, during breakfast at the mission-house, one of the missionaries, who acted as his tutor, introduced the boy to me as "our little chief." He is the son of the late king of Moskito's sister, Princess Victoria, the king having married a woman of mixed race, a creole. It is the law of the land that none but those of pure Moskito descent shall succeed to the chieftain-ship.

The little chief seemed to take a great fancy to me, generally accompanying me when I went for a stroll with my gun. He was about ten years of age, and appeared very intelligent. He lived at the mission-house, and was, I believe, well grounded in his studies.

Mr. and Mrs. Lündberg were very kind to me on this and on subsequent occasions when I stayed at Blewfields, insisting on my taking my meals with them, and showing me many other attentions, for which I shall ever feel grateful. When I was at Grey Town, Doctor Green, H.B.M. Consul, had very politely placed his empty house at my disposal, in consequence of which I took possession of a hammock that I found in one of the rooms.

The government of this country seemed in a very unsettled state. The Spaniards of Nicaragua had not recognized the succession, and contemplated making an attempt to take possession of the whole territory. Both the missionaries and the natives (whom I have much cause to esteem for the kindness shown me whilst I lived among them) are much distressed at England's having withdrawn her protection, and at the treaty made by the late Government, in which the best part of the territory was handed over to the Spanish states. In fact, I believe there is hardly a Woolwa Towaca, or any other of the pure aboriginal tribes, within the present boundaries of the territory. These Indians do not live in the swamps on the coast, but, in order to discover their settlements, it is necessary to ascend the rivers to a considerable distance, to where the banks are high. I believe the Spaniards would find it a difficult matter to force them from their homes, and drive them into the swamps which they have marked out for the Indians' territory.

The weather in the beginning of November was rather stormy, accompanied with some heavy rain, but this kept the air delightfully cool. The settlement of Blewfields is situated on a little peninsula in the lagoon, a fine sheet of water about twelve miles in length, and divided from the sea by a low, narrow slip, called Deer Island, from the number of these animals said to be on it. But they are now rarely seen, on account of the tangled state of the bush since the hurricane, which, as the natives say, quite " mashed up " the woods. Racoon, however, are still common among the broken mangroves on the lagoon side. There are many other smaller islands, the most interesting being Rama Cay, a settlement of christianized Indians, of the Rama tribe. Mr. Rhan, an Englishman, had also built a solitary house on Cassada Cay. It forms a conspicuous object on passing the bluff and entering the lagoon.

The population of Blewfields is of a very mixed character, springing probably from the origina Moskito race, and the two other races with which they held most friendly intercourse; namely, the runaway negroes from the West Indies, and the English creoles.

Many people in England, on hearing of this almost unknown country, are probably puzzled by the name *Moskito*, and possibly connect it in their minds with an unusual abundance of the well-known insect plague of the tropics. This is, however, a mistake, as they are by no means so numerous or so troublesome here as they are in other quarters, both of South America and the tropical countries. The name is a corruption of that belonging to the principal and most war-like tribe in this part of Central America. They term themselves Miskito or Moskitos, and were in times past exceedingly powerful and successful in resisting the inroads of Spain, with the assistance of the English.

Some time before I visited the country, the Moravian missionaries on the coast were distressed by the appearance of a Spanish priest from Nicaragua, who arrived with a canoe-load of tobacco, &c. It was the padre's custom to collect a number of Indians round him by presents of the much-prized weed, and to seize the opportunity for preaching to them in their own language (by no means a difficult one), and to baptize them indiscriminately. He soon left the country and went to Grey Town, and the only trace he left behind him was a few Christian names, the bearers of which seemed to be sensible of no change having come upon them, as they never could even have been taught the rudiments of the Christian verity.

While at Blewfields I was shown an imperfect skin of a very curious little animal, rather larger than a good-sized rat: the short, soft fur was of a greyish colour, and the tail short and flattened, not unlike that of the beaver; but the head and portions of the skin had been eaten away by ants and cock-

roaches, the pests of the country. This creature is
rarely seen, and then only at the water-side.

The Moravian mission at Blewfields seems to be
very prosperous; both the schools and the church
services seem to be very well attended on Sundays,
and on week-days also. It is a pretty sight to see
the people assembling for Sunday's religious teach-
ing, the women in their neat, clean cotton dresses
and white handkerchiefs; the men, for the generality, in white trousers and bright-coloured shirts,
though some appeared in European coats, which, in
my opinion, were not so much in character. The
church, like everything else, had been blown down
by the late hurricane, and was not yet rebuilt, so
that the services were held in the school-house.

A German colony was once established here,
and had cleared a large space of ground behind the
houses, but numbers had been carried off by disease,
and the rest are now scattered, some at Grey Town
and elsewhere, and some I afterwards met in
Nicaragua.

In my walks about the clearings I was struck by
the infinite variety of butterflies; but the only birds
often seen, besides the universal buzzard or John
Crow, were a few hawks, perched on the tall, bare,
eboe trees, which stood alone upright amidst the
general prostration, and a blackbird with a long tail
and a peculiar note *(Crotophaga sulcirostris)*, called
by the Indians " pequil," in imitation of its cry. I
observed also a species of finch *(Phenicotheanpis
Eubicordes)* which the creoles call " ground sparrow,"

hopping about among the low bushes near the houses. Many melodious notes, especially in the early morning, came from the guava thickets, but I was unable to obtain a good view of these shy minstrels.

CHAPTER II.

AT daybreak on the morning of the 5th of November, 1866, I started for Blewfields; or, as it is sometimes called, from the Indian tribe that inhabits its banks, the Woolwa River. It flows into the northern extremity of the lagoon by many mouths lined with mangrove thickets, now destroyed by the hurricane; in fact, many of these channels are completely closed up by trees that have fallen across them, and by an accumulation of drift.

For my journey, I had engaged a large pit-pan, or river canoe, and three men, at the following rates, as I did not know how long I should require them: Nash, head man, four shillings a day; his son, and a mixed Moskito man, called Teribio (who was going home to his Woolwa wives—he had two at Kissalala), at three shillings each; and the pit-pan one shilling. A lad also joined us to work a passage as far as Kissalala, the first Woolwa settlement, where his father, Hercules Temple, had a small trading-place.

We embarked, and took in provisions at a point opposite the mission-house, and here I got a cup of hot chocolate from my head man, which was, I thought, very much better than the sweetened stuff called chocolate in Europe. After this delay, the men dashed their paddles into the water in earnest, shot rapidly across the northern corner of the lagoon, and entered one of the many channels between the mangroves leading to the main river; after that, we went slower, as I wished to look out for birds to shoot and skin. When we had entered the main river, and cut a supply of sugar-cane, which here grows almost wild, and without which the inhabitants never think of going a day's journey, we set sail. The banks are very flat, the only elevations being one or two round hills standing a little way back on the left. The forest was a vast scene of destruction, nothing remaining after the hurricane but bleached stems and broken limbs of trees.

The Blewfields people have many plantations a short distance up the river, which they visit from time to time to fetch provisions, and to plant what they require for the next season. I saw many more people here in coming down, as it was then the season for burning the bush, and making new provision grounds, and also for grinding the sugar-cane. The birds most common were fish-hawks, which were very shy; king-fishers of various sizes; and small white and blue herons, or garlings, as the creoles call them.

We passed the night comfortably at a little bamboo-house on one of the plantations. Nash was very attentive in arranging some split bamboos, on which I might spread my blanket above the damp ground. Our supper was characteristic; and

I tasted several native dishes for the first time; the principal was a large iguana lizard, which was very good eating, with boiled cassada: its flesh was remarkably white and tender. Whilst I lay down after supper, smoking my pipe, the men amused themselves with talking about our sea-captains, and their fights with the Spaniards. Another favourite theme was the clever ruse played by the English Colonel Courcy, and his European and American officers, upon Walker's filibusters. While the latter were watching the river (the San Juan), little expecting an attack from the land side, where they thought themselves sufficiently protected by the thickness of the bush, Colonel Courcy cut his way through the woods to the rear of their works, and took them completely by surprise.

By dawn next morning we were again paddling up stream, under shadow of the bank. The river was like glass, and the only thing endowed with life to be seen through the slight mist that hung over the water, was the shadowy flight of a large grey crane we had disturbed at his early breakfast. On this morning I shot the first specimen of a handsome bird, called by the creoles "yellow tail," (*Ostinops Montezuma*). We had dinner at a plantation belonging to one of Nash's relations, and enjoyed another good meal of iguana and cassada. I also discovered that a heron which I had shot was very good eating. In the afternoon we met a small pit-pan with four Woolwa, two men and two boys, paddling down to Blewfields. Their skins were of a warm reddish-brown colour, and the two men had very fine faces: they kept their canoe alongside, whilst speaking with Nash, by a most graceful movement of the paddles. Not long after, we passed the last plantation of the so-called creoles, Nash pointing it out, and informing me that it belonged to him. I shot and skinned several birds on my way, but I had no idea, until a trial, of the difficulty of drying them, on account of the frequent showers.

We continued our course through part of the night up the river, which is very serpentine; and when we reached the mouth of the Rusewass, called, in Nicaragua, Mico, we secured the canoe to the boughs of a swamp-wood that hung over the water, as there is no good place here for camping on the banks, and had rather an uncomfortable nap. We started again before day-break, but stopped about breakfast-time to boil some water and make tea, on a flat rock under the banks, which began to rise to some height, while the river became swifter and more picturesque. It was still early when we arrived at Kissalala, where the river is a rapid stream, flowing between high banks; here we landed, and, after climbing the steep ascent, approached the first Indian lodge. Bowing my head, I stepped across the little trench, and passed under the low-hanging thatch. I found myself in what appeared quite another world of manners and customs, which made a strange impression upon me, so totally different was everything that I now saw from all my previous experiences of life. Since that time, I have learnt to feel quite as much at home in an Indian lodge as in any other place.

But to return to my first introduction to the Woolwa. The only notice apparently taken of my arrival at the lodge by its inmates, was to motion me to one of the low cedar stools, which were the principal part of their furniture, and then I was left to make myself at home. The women, in their decidedly light apparel, continued to busy themselves at the fires from time to time, stirring the contents of the large pots with long-handled wooden spoons, while the men went on tipping their arrows, carefully testing their straightness and balance by looking along them when held at arm's length. They relaxed somewhat, however, on receiving a present of tobacco, of which they are exceedingly fond, although they do not themselves cultivate the weed. Our conversation then turned on the "old times," when the territory was under English protection, and "man-a-wars" frequently visited the coast. They said that a captain once came up the river as far as their settlement, and described him as having taken a sketch of the lodge, with the women in the act of grinding corn. The man Teribio was given to talking of the "man-a-wars" men with whom he had been acquainted in those days, and he showed a great talent for imitation, by the thoroughly English way in which he used his fists. On one occasion, when some of his Woolwa guests became rather boisterous and quarrelsome, over a *mishla* feast he had provided, he settled differences by knocking two or three of them over right and left, in the most approved fashion of the P.R. This mode of proceeding gave him a great advantage over the Woolwa, who usually strike with their elbows. As a people, they are generally very peaceably disposed, seldom engaging in strife, although frequently intoxicated at their mishla feasts. On these occasions they generally amuse themselves with talking all together at the tops of their voices, when, of course, no one has the least chance of being heard. It seems to be a point of honour with them to give and then take a blow in turn, and not to try how often they can strike one another, as much as how long they can stand it. Although the elbowing is an awkward method, still the blows when delivered in this way, full upon the chest, have an ominously heavy sound.

It seems probable that the curious custom of flattening the forehead, which so largely prevails among the aboriginal American tribes, had its origin in a desire to increase the characteristic formation of the head, which they naturally must consider the highest type of beauty. The Woolwa, however, do not practise this fashion to the same extent as the other Indian tribes; and the hair, which is worn hanging to the eyebrows in front, grows so thickly that a casual observer would hardly notice any peculiar flatness of the skull. It is interesting to take note of the various effects pro-

duced in the figures of the different races by their
distinctive modes of life.   Among the Woolwa,
there is a large development about the arms and
chest, whilst the lower parts of the body are often
inclined to be squat.   This is, doubtless, caused by
their habit of spending much of their time in paddling, poling, and hauling their pit-pans up the
creeks and rapid rivers.   Indeed, they are as essen-
tially canoe-men as the civilized Indians in the
district of Matagalpa are pedestrians.   The Woolwa
places of burial are always in the vicinity of the
river-banks, and are marked by a large, thatched
shed, similar in its construction to the lodges
inhabited by these Indians.   This is built over the
spot of interment, and the whole is sedulously kept
clear of bush-growth.

This tribe has a singular mode of playing with
staves or short poles, which they grasp in the
middle, and then, standing opposite each other,
hold them at arms'-length, and strike each end
alternately together with all their force.   The oppo-
nents are matched in pairs, and in appearance it
rather reminds one of the old English quarter-staff
play.   The object of the game is to see which can
keep up longest the continual strain upon the
muscles of the arm, and ultimately strike the staff
from the hand of the other.   It is strange what a
wide spread the childish game of " scratch-cradle "
has.   The Soumoo Indians appeared to me to carry
it into far more complicated passages than we do.
The Soumoo seemed frequently troubled with pains

to face page 160.

SOUMOO OR WOOLWA INDIANS, DESCENDING A RAPID IN A PIT-PAN
CENTRAL AMERICA

in the head and limbs.   For the former ailment
they were in the habit of tying a cord tightly
round the head; and for the latter, they flogged
their limbs with a kind of nettle, until the skin was
raised in bumps.

The Woolwa have many strange customs attend-
ant upon their coming of age.   The young men
have many physical ordeals to undergo before they
are fully entitled to the privileges of man's estate.
Among others, they have to bear heavy blows on
the back, given with the elbows.   This, although
well enough for the strong, must press heavily on
the weak.   The rest are of a similar character,
all being apparently dealt with the intention of
ascertaining what amount of physical suffering they
can endure.   It seems probable that these customs
are but the remnants of more useful exercises, calculated to strengthen and educate their bodies in
the art of war, at a time when they were a more
numerous and warlike people, and also to teach
them that fortitude which is so highly esteemed
among all the Indian tribes.   The Woolwa must
also be expert swimmers, as they usually bathe
several times in the day; but an opportunity rarely
occurred of observing them in the water, for, when
men or women wished to bathe, they usually
stepped into one of the canoes which were moored
at the landing-place, and dropped down the stream
to some secluded spot, where they could go through
their ablutions in privacy.

I am quite certain that, during my sojourn

M

among them, these people enjoyed many a good
joke at my expense, from the manner in which they
would look at one another, say something in their
own language, and laugh heartily. No doubt other
travellers have found themselves in the same not
very agreeable position of affording amusement to,
if not successful in " astonishing the natives."

The Woolwa, or Soumoo, as they always call
themselves, and prefer to be so styled, have no chief
of their own ; therefore, when any difficulties arise
(which is not often the case), they go to Blewfields
to settle them. They have no villages of any size,
but live in lodges, grouped two or three together,
and scattered at intervals along the main river and
its tributaries. These lodges have no walls, but are
open on all sides, which I did not find uncomfortable in that hot climate, as they are sufficiently
sheltered from the rain by the palm-leaf thatch,
arranged to hang so low that one has to stoop
in entering. This thatch has a very neat appearance, especially from the inside, which is usually
decorated with the lower jaw-bones of the peccary
and warry, or wild hog, &c., and also the bleached
skulls of large fish. Sometimes there are stages
made of split bamboo, for storing away dry maize
and other things; and bows and arrows, the only
weapons now in use among them of their own
manufacture, are stuck into the binding of the thatch.
The rest of the furniture consists of one or two
wretched old guns, obtained from the traders in
exchange for their canoes, india-rubber, and other

articles, an axe, and a few rusty machetes; stones
for grinding corn, earthenware pots of their own
making, decidedly picturesque in shape, in which
they cook their food, perhaps also a cast-iron one,
bartered from the traders; some odd-looking little
bags suspended under the eaves; pieces of native
cloth hanging on the supporting beams; a cradle,
with the dried claws of crabs and other things
attached to it, that make a strange rattling noise
when it is rocked—a sound often banishing the
stillness of night. Four families generally inhabit
each lodge, each having their fire in one of the
corners, at which they do their own cooking, and
sit round chatting. These lodges are usually surrounded by a number of the most miserable-looking
curs imaginable, constantly on the watch for what
they can steal. The Indians are very fond of
taming wild animals for pets, and you seldom stop
at a village where you do not see parrots, parroquets, monkeys, or tame warry and peccary. At
one place I saw a little boy running about with
a tame otter, here called " water-dog."

Having given this sketch of the manners and
customs of the Woolwa Indians, I will now proceed
with our journey. The following day we went to the
next Woolwa settlement, a small village of about five
families, and on our way we passed two rapids, at
one of which we had to empty the pit-pan and haul
it over the rocks. The Indians of this settlement
were very good-natured, as I always found them, in
helping us to carry the things up the steep bank to

their lodges. The name of the eldest man was
Kennedy, and, in fact, nearly all were known by
some English cognomen, acquired in their visits to
Blewfields.

In the afternoon all the men came and sat
down, conversing with Nash, and smoking tobacco
I had given them out of one clay pipe (obtained
from the traders), which they passed from one to
the other. They seemed interested when I told
them that other Indians, living far away to the
north, in quite a different sort of country, smoked
in the same fashion, and were most civil and kind,
knocking down the ripest oranges from the trees,
and doing many other little things to please.

I passed the night here, and next morning
started again up the river. I saw very few birds
at this place; indeed, animal life seemed very
scarce all along the river, whether from the effects
of the gale or not, I cannot say. We proceeded up
the river the rest of that day, but the weather
changing for the worse, I thought it better to turn
back to the first Woolwa settlement, where I proposed to remain until it became finer. Accordingly,
next morning, we dropped down the river as far
as Kissalala. As I expected to pass some time here,
I settled with the head man and his companions;
so the trio returned in the pit-pan to Blewfields.
Left alone in my large lodge, I soon found that the
life of a solitary traveller is not an idle one, for
having to be at once master and man, renders his
position no sinecure. That afternoon I made my

first attempt at cooking my own dinner, which consisted of rice, boiled with a few drops of cocoa-nut oil, and I flatter myself that it was a decided success. I like the flavour of the cocoa-nut better than that of the milk we use at home. A very welcome addition to my cassada and plantain were the birds I often shot in the morning,—some of them, especially the water-birds, such as herons and rails, being very fat and rich. I always concluded my repast with a good strong cup of tea, brewed in the Australian fashion, and then came the pipe, never a greater source of enjoyment than on such an occasion. Nash's son, who was staying here for a short time, often came into my lodge, and insisted, out of pure good-nature, on cooking my meals : he cooked rice deliciously with fresh cocoa-nut oil, but he was much too extravagant with the materials, for, having a long road before me, I was obliged to be economical. By the time I had been three days in my new home, I became used to this style of living. I lay down and rose up again with the sun, though occasionally I skinned a bird by the light of my bull's-eye, which I may remark, *en passant*, is a very bad sort of lamp for travellers. Game appeared to be rather scarce, and this, like every other ground of complaint, was attributed to last year's hurricane. Mosquitos were not numerous—at Blewfields hardly any are to be seen ; but it was a long time before I became used to the ants, crickets, and cockroaches, whose crawling, scampering, and buzzing kept me

awake for many a long hour; in fact, nothing can
be more disagreeable than the habit the large flying
cockroaches have of alighting on one's hair and
getting their legs entangled: there is a consolation
in knowing that they are easily killed. I used to
find them dead around my blanket in the morning,
from my having filliped them away during the
night. One morning, after being very much disturbed by the creeping of ants, I discovered that
a colony of these little creatures had been employed
in shifting their quarters to the blanket which had
been rolled up for my pillow, and had already
stored away a quantity of eggs in the folds. After
this I always took the precaution of spreading my
blanket close beside my fire, and was, for the
future, very little troubled.

While staying at Kissalala, I shot a number of
small birds round the edges of a patch of ripe
maize, or Indian-corn. At this season, also,
numbers of beautiful glossy blackbirds, known
to naturalists as *Cassidix oryzivora,* and called by
the Woolwa *Mook-ris* (black-bird), used to resort
here to eat the ripening corn. These, together
with the pigeons which I shot in the trumpet-trees,
made a very pleasant variation in my daily fare.
There were also three very beautiful birds of the
oriole kind, clothed in orange, yellow, and black,
in different proportions, and called by the creoles
banana-birds. They enlivened the early mornings with their rich mellow notes, and often sang
during the day, especially when the sun came out

after a shower. I heard also, from time to time, strange notes in the recesses of the woods, but the bush was so interlined, by fallen trees being covered and matted together by convolvuli and other creeping plants, that it was impossible to penetrate into the woods for any distance, for the purpose of shooting the musicians.

Among the most attractive of the small birds around the clearing, were two kind of finches: *Kamphocælus passerinii*, and *Kamphocælus sanguinolentus*. In these species, the back above the tail was adorned with a most brilliant patch of scarlet feathers; and the only noticeable difference between them was, that in the smaller the whole of the plumage, with the exception of the patch of feathers already mentioned, was of the deepest velvety black, while the larger variety had a band of red across the head. They were very lively in their movements, looking like a gleam of fire darting amidst the bright green foliages of the shrubs. The bird called by the Indians *Pegua* resembling the Mexican cuckoo, is a common bird on this river; and its cry, from which the name is taken, uttered from time to time, as if in a tone of discontent, can be heard on all sides of the forests. The natives say that they utter this cry when unsuccessful in their search after insects.

It is an exceedingly difficult bird to skin, and when hit, even with the smallest shot, the feathers usually fly off in a cloud. The large yellow-headed parrots and the macaws used to fly in pairs, high over-head, every morning, with loud cries, returning again in the evening to roost in the eboe trees at the back of the settlement, where they would chatter whilst the last rays of the setting sun appeared above the forest trees: then the fires in the lodges would blaze up brighter, and the Indians took their last meal, and afterwards spread their sheets of bark cloth for the night. The strange cry of a species of goat-sucker, "Who-a-you? who, who, who-a-you?" would come from the neighbouring corn patch, to be answered on the other side of the river, and then gradually would die away in the distance down the stream, and, if the night were very bright, the loud rattle of a kind of crake could be heard among the reedy sedges on the water side. In the early part of the night in wet weather the noise made by the frogs was almost deafening; but later on no sound, except the occasional hoot of an owl, broke the universal stillness.

I know of nothing so suggestive of reflection, tinged with a wholesome sadness, as thus to find oneself alone in the pathless wilderness, associating with a race utterly strange, not only in habits, but in appearance, and whose language, foreign to the ear, conveys no intelligence to the mind. Thus situated, one feels truly brought face to face with the Great First Cause of all, and to appreciate the full benefit of prayer. New beauties and new meanings are seen and felt in Holy Writ where they had never been recognized before. As evening falls, the mellow light of the setting sun forms a

glorious background to the giants of the forest; unfamiliar sounds arise from its recesses, and a great
calm steals over the senses, not of sorrow or of joy,
but peace, that seems to realize the words of the
inspired writer, " They that weep as though they
wept not, they that rejoice as though they rejoiced
not, for the fashion of this world passeth away."

But again the world closes round ; one is drawn
back into its strife and turmoil, and all the newly-
awakened solemn feeling is remembered only as a
dream or bygone sensation.

On the 14th of November the rains seemed regularly to have set in ; we had heavy showers every
day, and I did not anticipate fine weather before
Christmas, after which I was informed it would
improve. The heaviest rains usually fall in June
or July, accompanied by much thunder and lightning. The river now began to rise rapidly, and I
could get no fish. The Indians often brought me a
large bunch of green plantains, for which I gave
them some six strings of beads; roasted in the
embers, these were very good indeed. I have never
tasted them anywhere cooked so well as the Indian
women do them : they take great pains with them,
turning them constantly, till, when they are done
and broken open, the inside is quite soft, and smells
like new bread. Cassada also is very nice when
fresh, but as it does not keep, it is of little use for
travelling. I was surprised to find that these
Indians do not make flour or cakes from it, as they
do elsewhere, but always either bake it in the embers

or boil it with their meat.   There is another root,
coco, of which they seem very fond, but I could
never bring myself to like it: they use it to thicken
soup.

It poured in torrents on the night of the 15th,
and the roof of my lodge, which was very old, began
to get leaky; but the next day was fine, though the
river was high, and the current swift and muddy.
During the rains the Blewfields river rises with
astonishing rapidity, and the current becomes very
strong and turbid, bearing along its logs and trees.
At Kissalala I have frequently watched the blue
garlings *(Ardea cœubea)* perched on the floating
masses, on which they allow themselves to be swept
down the stream to a certain distance, when they
take wing and ascend until they meet another uprooted tree, on which to be rafted down the same
distance, and so on, repeating the process again and
again.

Sometimes, while passing under the trees in a
canoe, I have noticed notches cut out on the trunks
at a great height, which I was told had been made
on the occasion of a flood.   At such seasons the
upper part of the river becomes impassable, boiling
and seething over the huge boulders that block up
its bed.   Several species of stinging flies used to
frequent the lodge, and run about over the thatch, in
search of the enormous hairy spiders that skulk
away in the crevices.   It was interesting to observe
how easily these slender and elegantly-shaped wasps
would   overcome   the   really   formidable-looking

spiders by pouncing upon them and stinging them.
The strong legs of the spider immediately release
their hold and become powerless, and they drop
down at once upon the earthen floor, followed by
the active fly, which then, after absenting itself for
a brief space, apparently to explore, commences to
drag home his gigantic prey, many sizes larger than
himself. The lair of the wasp is generally a hole
burrowed in the earth, but one that I frequently
watched had taken up his abode in the hollow end
of one of the bamboo rafters of the lodge. The
natives say that the sting of this insect is very
severe, as I was led to suppose from the deadly
effect it had upon the spider so much larger than
itself.

A species of ant is a useful but somewhat
troublesome insect, which invariably congregates
in numbers. They are in the habit of marching
in regular columns, but when they meet with what
seems likely to make a good hunting-ground,—an
Indian lodge, or a fallen tree, for instance,—they
spread out in search of prey. Cockroaches, immense
spiders, and other insects, and even animals, beat
a speedy retreat at their approach; but those un-
luckily cut off from safety by the converging
detachments, are instantly covered by swarms of
ants, and are bitten and stung to death. These
ants have a peculiar aversion to wet, which seems
rather strange, as their native atmosphere is exceedingly damp. When the Indians want to turn aside
the tide of invasion from their houses, they take

advantage of this peculiarity, by simply squirting
mouthfuls of water at the head of the column.
Many tales are told of sick persons having been
much injured, as they lay helplessly in their hammocks, by these ferocious legions.   There are also
several species of solitary ants, sometimes of large
size, generally seen on the stems and leaves of
trees.   The bite of some of these insects is much
feared by the natives.   On one occasion, a fire-ant,
as it is designated, alighted on the front of my
flannel shirt, and Temple filliped it off, with the
remark that had it bitten me it would most probably have caused a severe fever.   Cockroaches, of
various sizes and shapes, are one of the greatest
pests of the country, and stow themselves away by
myriads in every possible hiding-place: in your
knapsack, behind the lining of your hat, even in
the very pockets of your coat, whilst flattened and
dried specimens are to be met with in almost every
other leaf of your pocket-book.   After having been
some time on this river, I was compelled to give
up the luxury of brushing my hair, as every morning I found that simply tapping the back of the
brush over a fire caused myriads of minute cockroaches to fall in showers from the hairs, where
they had comfortably ensconced themselves during
the night, doubtless finding the position a covert
much to their taste.   The odour peculiar to the
cockroach kind, with which the brush became
impregnated, was so unendurable, that I had to
content myself with only passing a comb through

my hair. When a heavy shower fell, these pests
might be seen swarming into the lodge from the
bush beyond, in order to secure the shelter of the
thatch. It was perfectly useless to wage war upon
such a host. It was almost impossible to cover up
the provisions over-night sufficiently to prevent
them from devouring a large portion, and rendering
the remainder uneatable, from being so much
crawled over. They were by no means particular
in their tastes, and especially relished the paper
labels off bottles.

There were visitors of importance staying at
Teribio's lodge, on the other side of the place, but
I did not know who they were until they had
returned to Blewfields, although they were sometimes kind enough to send me a nice mess of iguana,
with plantain, roast, boiled, and mashed. Teribio
afterwards told me that they were the old queen,
Princess Victoria, and the creole wife of the late
king.

I often went out shooting in a small pit-pan
with a handsome young fellow, named Freshwater,
who was the eldest son of a decidedly plain, but
very good-tempered little woman. On such expeditions an Indian is more useful than any other man.
It was wonderful how noiselessly Freshwater would
drive the canoe through the water. When he saw
anything he would give a low grunt, point it out,
and wait quietly till I observed it, and then follow
the direction indicated by a motion of the hand as
silently as possible. He always carried with him

his bow and fish-arrows, to strike the fish when he saw them lying under the sunken logs in clear water. My little English terrier, Jack, was now becoming quite expert in finding birds that fell in the thickets, but I was obliged to keep quite close to him to prevent his tearing them. Of this trick I could not break him.

Time slipped away quickly in the various occupations of shooting and skinning birds, chopping wood, and cooking. A little Indian boy, brother to Freshwater, attracted probably by the handful of sugar I sometimes gave him, used to visit me during my culinary operations: the confidential air he assumed, while he stood on the opposite side of the fire, was very amusing. He used to chatter in the strange Soumoo language, evidently imagining that I understood all he said, and would nod and grunt his assent in a most satisfied manner when I replied to him in English. The little fellow seemed hardly to have recovered from the flattening of the forehead to which the tribe subject their infants; his eyes were still very prominent, and had a peculiar staring expression, which I have noticed in the eyes of most Indian boys. He seemed to have a great affection for animals, and, when running about, had usually some little pet attached to the end of a string—sometimes it was a large kind of mole-cricket, which every now and then buried itself in the ground, only to be dragged out again by the cotton-thread tied to one of its stout legs; at other times a little bright-eyed iguana, or a *karkee*, a little creature resembling a rabbit. On one occasion his foot was severely bitten by a young otter with which he was racing about. Altogether, this little Indian was one of the oddest children I have ever met with. I shall never forget the look of terror depicted in his face when some of the girls pretended to drag him towards the large black morose howling monkey, *Almook*, as they call him, which was tied in a corner of the lodge, during the time I possessed him. This howler seemed to be held before the eyes of the dusky youngsters of the settlement in much the same manner as the unfortunate police at home are by nursemaids, to frighten their charges into obedience.

When preparation was being made for hunting expeditions, the peculiar noise made by the women in grinding their maize on rough stones often continued far into the night; and as I lay awake, I was forcibly reminded of the "sound of the grinding" mentioned in Scripture. The grinding-stone used by the Woolwa is also used by the Nicaraguans, who call it *metlate*, evidently an aboriginal word. After the maize has been thus prepared, it has the consistency of thick paste, and is taken by travelling parties in their canoes, folded up in *waha*, or banana leaves: the former is generally selected for this purpose, as it is much tougher, and not so liable to split. When kept more than a day or two, the smell of this paste becomes sour to a sickening degree; and when requiring it for a journey, I always made a point of having it stowed away as far back as

possible in the stern of the canoe. The Indians
usually halted at mid-day, in some shady spot, in
order to enjoy it; on which occasions it was prepared for drinking by simply squeezing and mixing
a handful or two of the paste in a calabash of water:
at other times it is rolled in a leaf, and baked in the
embers, which mode of cooking renders it by no
means unpalatable. One thing which struck me
much was to hear the little Indian children
addressing their mothers precisely as European
children do, and running to them with the familiar
cry of "Mum-ma."

A few months before I came to Kissalala, a dark
tragedy must, from the rumours I gathered, have
been enacted up the Rusewass tributary. The reports
I had heard were contradictory, but the facts seem
to have been much as follows:—Some Spaniards, from
the state of Honduras, I believe, had ascended the
Rusewass for some distance, and had built houses
and established themselves in order to collect india-
rubber. The Woolwa of that river then demanded
payment for the clearings and maize plantations
that the Spaniards had made. This being refused, a
dispute arose, in which the Indians, having threatened
the others with their arrows, had one of their number
wounded with a machete-cut; on which they with-
drew, but some time after, they planned a night
attack upon their enemies. Surrounding the place
when least expected, they are said to have "clubbed"
every one, leaving none alive to tell the tale. It is
probable that should the Nicaraguans attempt to

occupy the country, there would be many such instances, as the Woolwa, and other inland tribes, would be easily influenced by the more warlike Moskitos.

The Indian dogs of this country are the most atrocious-looking curs that can be possibly imagined. Their shapes are often extraordinary, but never graceful; and one especially struck me as remarkably hideous. It was a rusty-black brute; all the bones of its long, thin body appeared distinctly beneath the skin; and this body was set on legs so short, that it scarcely kept clear of the ground. The fore legs were so bent in, that the beast walked more upon the joint than on the foot, which was armed with formidable claws like a bear; and, to complete its repulsiveness, it had a most villainous leer in its bluish-grey eyes, as it would look up in my face and snarl when disturbed in the act of thieving. The Indians seem rarely to feed their dogs, and, therefore, they are continually prowling about, to pick up a living by what they can manage to steal. On this account, I was the more surprised that the *miril*, or Indian women, should take such a fancy to my little English terrier, Jack, as to feed and pet him; but I suppose the honest, good-tempered expression in his intelligent brown eyes contrasted favourably with the sneaking looks of their own curs.

Nothing appeared so much to astonish and amuse the Indians as to look through my powerful telescope. Sometimes I set the focus for the other

N

side of the river, in the direction of some cattle belonging to Teribio, which had strayed away; by their great weight they had broken through the jungle, and were now inhabiting a sort of yard amid the impenetrable thickets on the hill-side, formed by the fallen trees that were covered and interlaced by bushes, vines, and convolvuli. It seemed quite a marvel to the Indians to be able to discern the slightest movements of their sleek, well-fed beasts, although it was still out of their power to secure them.

A curious native drink is made here from ripe Indian corn, parched over the fire in an old pot; then ground fine, and mixed in water sweetened with syrup. It is cooling and refreshing, but the dry particles occasion a tickling sensation in the throat; and unless it be continually stirred during the act of swallowing, one gets little but sugar and water. When the Woolwa women present the calabash, or drinking-bowl, to their guests, the offer is invariably proffered three times before the same ceremony is gone through with the next person.

On the 19th I received an agreeable surprise, in the shape of letters from home. Mr. Lündberg kindly forwarded them up the river by some Woolwa who were returning to their settlement. I immediately left the bird I was skinning, that I might get my dinner early, and be able to enjoy the luxury of reading my letters over a cup of tea and a pipe.

I usually went out shooting on the river in the

mornings. Some of the birds are very hard to kill
with small shot; and of this I had proof one afternoon, for I fired four into a fish-hawk on a tree,
knocking out a cloud of feathers at each shot;
but after all, he managed to escape into the neighbouring thickets. One of the men ,brought me
a snowy-white falcon, which he had slightly
wounded in the shoulder, whilst up the Rusewass
looking for rubber. It was one of the most beautiful creatures I had ever seen; its feathers being of
the purest white, with the exception of the extreme
tips of the wing feathers, which were black: there
was a bar of the same across the tail. The eyes
were large, dark, and of a wonderful brilliancy. I
kept him tied to an eboe-log in the corner of the
lodge, and do what I would, whenever I looked
towards him, I found his lustrous eyes following
me with an imperturbable stare. I was extremely
sorry to find that his wound was mortifying; so, to
give the noble bird a chance of life, I cut the cord
that bound him, and he stole away into the thickets.

The kind of canoe principally used on the
coast and the deeper parts of the river is called by
the natives a *dory*. Having a keel, it requires a
greater draught of water than the pit-pan, which is
for the rapid and shallow parts of the river, since it
is nearly flat-bottomed and square at both ends,
which project above the water. Both these canoes
are cut out of solid cedar-trees. The pit-pan is very
thick-bottomed, which renders it capable of sus-
taining very rough usage in hauling it over the

rocks at the portages; it has also a square mortice-hole cut in the stem, through which a pole may be thrust when it is necessary to moor it. The distinguished visitors who had been staying here returned about this time to Blewfields. The old queen, grandmother to the little chief William, was evidently the most important personage; she was a very fine old woman, and appeared very vigorous for her years as she marched down the pathway leading from the steep bank to the canoe, where she took her place in the middle. The widow of the late king took a forward paddle, and a young Moskito man the stern; the rest of the space was occupied by the children, the plantains, and other provisions. The "old queen," as the people call her, was of the pure Moskito race, with fine, high features, but much darker in colour than most of the Woolwa.

I now engaged a thick-set old Woolwa, with whom I went a long way down the river in quest of birds. I shot the first specimen I had seen of the large handsome bittern (*Tigrisoma cabanisi*), which the Indians call *woukee*. I often afterwards made a meal of them, as I found them fat and well-flavoured. They are not numerous on the river. I shot a great many, at a later period, in the creeks on the coast to the north of Pearl Cay lagoon. In the evenings the men of the settlement usually returned with a heavy load of dry logs for fire-wood, procured from the dry drift-wood on the banks of the river, which they split into suitable lengths for burning. They used the axe exceedingly well, with great strength and gracefulness. The women then made up the fire and prepared the meal, which is brought to the men as they sit and talk. When sleeping in the Indian lodges, I used constantly to hear the peculiar slapping noise, indicating an onslaught upon the swarming insect plagues that settled on the skins of the recumbent Indians; these sounds becoming more frequent just before dawn, when the sand-flies turned out in force, at which time the men would get up, make the fire, and have a chat before they lay down again. Old Temple, too, would rise from his hammock, light his half-smoked pipe (he never smoked his pipe out at one sitting) with a brand from the fire, and then turn in again. On the 25th my solitude was somewhat broken in upon by the arrival of a *dory* from Blewfields, full of creoles, among whom was a trader, by name Hercules Temple, to whom I have alluded before; the other men he had engaged to collect india-rubber. These people gave unmistakable signs of their African origin; Temple fiddling away a good part of the night, whilst his son accompanied him on the top of an empty barrel, was a contrast to the quiet of the Indian part of the encampment. The new-comers, however, were very pleasant neighbours, and most obliging. Temple himself was nearly black, with crisp hair, like many of the Blewfields creoles; he assured me that his mother was an Indian woman of the Toongla tribe. The morning after his arrival he gave me a piece of

bread, which proved quite a treat, after having been
without it for such a length of time. On the morning of the 26th, shortly after sun-rise, the whole of
the population went off like a flock of birds, some
up and some down the river, leaving the Blewfields
trader, his son, and myself alone in the place—
perhaps they did not much like their noisy visitors!
These Indians have so little baggage, that before
you know of their intention to set out on an expedition they have already made the start. During
their absence I went up the little creek which flows
into the river just below the settlement, and shot a
strange-looking owl *(Syrinum perspicillatum)*. I
also made my first attempt at paddling, with which
I soon became familiar in the course of the different
expeditions with Temple, from whom I had purchased a paddle.

The miserable curs hanging about the place gave
me much annoyance, by tearing down the bird-skins
that were put out for drying in the sun, so that often
when I came in from shooting I had nothing left of
the best skins save a few scattered feathers. Temple
did me many little kindnesses; sometimes, when he
saw that I had been cooking nothing but plantain
for dinner, either a nice dish of iguana, a cake of
bread, or a calabash of ripe plantain "*pop*," which
I found very good when sweetened with sugar.

The last week in November was very fine, with
a continuance of dry northerly winds. Temple often
came into the lodge I occupied to have a chat,
seating himself on one of my boxes; he was very

much interested in my little sketch-book containing figures of Indians of different tribes, and was never tired of talking about it to the Woolwa, causing much merriment by his explanation of the differences of costume, &c., which he had gathered from me by questioning in a very self-complacent, dictatorial manner. I often gave him the monkeys and birds I had shot; in return for which he would send me some mess for dinner, much better cooked than what I could manage myself.

What struck me most about these creoles was the respectful way in which the son always addressed the father, invariably affixing " sir " to his answers.

The Sunday in a foreign land is a day which, in all its loneliness, always brings to a traveller thoughts of home. In this retired spot, with no countryman near, I appreciated more fully than I had ever done before the beautiful prayers and collects of our Church, especially when those for the first Sunday in Advent came round, bringing, as they did, old times very forcibly to my mind.

After this fine weather, the river had much subsided, enabling me to catch many good fish, and thus to supply myself with a little variety in the way of fare. There is one fish that I feel sure would afford good sport to the fly-fisher; the Woolwa are very expert in taking it with a hook, baited with a green grasshopper, which they catch in the long grass on the bank; this they tie to the end of a very long line attached to a switch rod, which they flick and cast about over the water, and letting it sink, presently raise it with a peculiar movement of the wrist. They seem to know the exact spot where the fish lie, and dexterously send the bait a long distance under the over-hanging branches of the swamp-wood bushes that shadow the deep still pools.

There was a very good place for bathing just above the settlement, from which it was concealed by a thick swamp-wood tree; beneath which was a piece of rock, from whence I took headers. Some of the evenings were so calm and mild, the river being like a mirror, that a swim was delightful. The first time I struck out into the stream, the Indians all clustered on the high sloping bank, seemingly quite amazed to see that a white man could swim. Temple increased their wonder by telling them that I could do much more, but that I was rather sick that night, and expatiated upon the difference between mine and the native stroke. Alligators, although numerous, are not at all formidable in this river, except to the dogs, of which they take a great many. The Indians here lost every one of theirs during my stay. They are generally taken while swimming on shore to chase the *karkee wastusa*, or Indian rabbit, into the hollow trees, where it is transfixed by the arrows of the Indians. I expected to have fallen in with the Honduras turkey; but after much inquiry, came to the conclusion that it did not come so far south, although Temple led me to suppose that it did, by his description of feathers worn by the Indians from the savannahs high up- the river. These savannahs, however, I discovered afterwards only to have existed in the old man's imagination, the whole course of the river lying through forest.

On the 1st of December I paddled down below the rapids with Hercules Temple and his son. The latter was a handsome, saucy lad, with a smooth black skin, and a bushy head of hair. He, like his father, could shoot a fish or iguana with an arrow as well as any Soumoo. There were not many birds to be seen—but we killed two monkeys; a very welcome addition to our provisions, as we had been for some time short of meat, and all the Indians were absent. The thicket which had sprung up in the shattered woods was in such a tangled state that we were obliged to cut a road with a machete (a weapon that Temple knew well how to use) up to the tree where we had observed our game. One of the monkeys, shot with an Eley's cartridge (of which I should advise every traveller to take a good supply), hung from the bough long after it was dead. We began to despair of our dinner. At last, however, Melville (Temple's son) succeeded in climbing, and shaking the animal down. It was here that I noticed for the first time the kind of bush-rope which gives water when broken. I snapped one in making an abortive attempt to mount the tree: a small stream of pure water issued immediately from the fracture. Temple told me that men who have been accidentally "bushed" for days have found great relief from its use.

Three kinds of monkeys are usually seen in these forests. The monkey that we had caught was called by the creoles the red monkey; it has four fingers on the hands, and is the variety that is generally eaten. Besides this, there is a smaller kind; black, with a white face; and the third is the black howler, which utters a cry not unlike a tiger's howl, and is especially vociferous before rain. The first time I heard it I was puzzled to imagine what the noise could possibly be, and my terrier, Jack, was even more bewildered than myself. The Indians who were paddling the canoe with me pointed to the large limb of a silk cotton tree that stretched over the water, and there I saw the howler, lying supinely along the branch, with only his black-bearded face visible. We did not disturb him, but passed on under the bough, while he followed us listlessly with his eyes.

On the 4th, some of the Indians returned from their hunting and fishing, and I was able to get a fresh supply of cassada and plantain, of which, for the last day or two, I had been quite out of stock. I should, without this timely arrival, have been obliged to live entirely upon the few birds and fish I had been able to obtain.

About this time I thought of going down the river to the Rama mouth, and ascending that branch in quest of birds; as here they were scarce, and

I desired to make up a case to send off by the December or January mail from Grey Town, before I went further into the interior. One day, while paddling up the river with a big Indian named Jackson, we passed under the tree in which I had seen the howler on a former occasion; but this time as we were passing under the limb where he still lived, he made such mocking grimaces, and hooted us so, that I sent a shot at the face, taking him fair in the forehead, with one of Eley's wire cartridges. At that distance it went almost like a bullet, knocking him head over heels with a splash into the water beneath, where he disappeared. In another minute he came to the surface, and struck out gallantly for the bank, but we intercepted him, and succeeded in getting him into the canoe. He again slipped into the water on the other side, and swam for shore, but we caught him, and this time I secured him with my leather shot-belt, and so got him home. The creoles call this monkey "baboon," the Indians, *almook* (old man), and they seem to consider him a sort of wild man of the woods, or rather a devil, and refuse to touch or even to approach him. I brought my captive into my lodge, as he did not seem to be severely hurt, the small shot not having penetrated the thick skull; and next day he ate some banana I offered him. He had a beautiful coat of the richest brown, deepening almost to black on the back, and a long black beard. These baboons, being very morose in temper, are not kept as pets; but I thought it might have been acceptable in the Zoological Society's Gardens.

We had had lately many heavy squalls of wind and rain, and on the 8th a violent storm swept through my sideless lodge, thoroughly wetting the skins I had dried, and carrying away with it various articles. The old roof bent in such a threatening manner that both Temple, who was with me, and I thought it must come down, and accordingly I ran to the windward side to get clear of its fall; but the squall passed over as suddenly as it came. I was obliged to remove my fire to another side of the lodge. Many of the patchings of *waha* leaves had been blown off the corner where it had hitherto remained, and the rain came in while I was lying by the fire-side at night, wetting my blanket to a very unpleasant degree. The river again rose very much, and large pieces of timber floated past. I was not now in want of anything to eat, as the women constantly brought me in plantain, Indian corn, or messes of meat; for which I gave them beads. The only thing I missed very much was butter, but I found that fresh cocoa-nut oil was by no means a bad substitute when it could be obtained; without something of the sort the corn and plantain are rather insipid. Muscovy ducks are not common, but I occasionally met with them, once shooting one at the settlement, which Temple had seen pitch among the bushes on the other side of the river, in some standing water that had been left by the flood. Teribio, who had now returned from the upper part of the river, gave an excellent account of the forest there, saying that it had not been "mashed up" by the hurricane.

A grand *mishla* feast was now about to take place, and the women were busy preparing the drink for some days beforehand: this is a very disgusting process, but is, I believe, connected with their religion (what that may be I do not know). These feasts, however, are carried on with a certain amount of decorum, different from the jolliness with which they hold carousals with drinks made from banana, sugar-cane, &c. *Mishla* is the general name for all kinds of drinks; but unless some other name is added to it, it is supposed to mean what is made from cassada. When the Indians intend to give one of these feasts or ceremonies, the whole community club together, and collect a large quantity of the cassada root, which the young women then commence chewing, spitting it afterwards into an earthen pot. When their jaws get so tired that they are obliged to desist, they boil the remainder, and, after mixing the whole, let it stand for a day or two, until it has fermented, keeping it stirred and skimmed. People are invited to come from a great distance to attend these festivals, on which occasions they are to be seen in their full costume of paint, feathers, and beads. Some wear a coronet made of the curly head-feathers of the curassow, which often looks very tasteful; also a cord round the upper part of the arm, from which flutter feathers of the macaw and downy owl, and the yellow tail-feathers of the *Osti-nops Montezuma*. The men decorate their necks with small opaque beads, procured from the Blew-fields traders, and worked by themselves into long pendent bands, often of very pretty patterns; these hang down in front of the body, and tassels of white beads fastened to a broad collar of similar work to the bands, depend from the back. The "*toumoo*," or "*pulpra*," as the Moskitos call it, is a cloth worn by the men round the waist, the ends of which hang down between the legs, generally below the knees, and with some of the young dandies reaches to the ground. This "*toumoo*," like the sheeting in which they wrap themselves at night, is made of the bark of a tree, beaten out by the women on a smooth log, with a mallet shaped like a club: there are grooves in this, which give it a texture and the appearance of a mesh. They are also made some-times of a very stout and handsome cotton material, dyed in many colours, and woven into tasteful devices, occasionally mixed with the down and feathers of birds. These do not seem to be much in present use, probably from the time and labour expended in the manufacture. The women, on full-dress occasions, wear a great quantity of beads round the neck; but, unlike the men, they do not work them into designs, only putting on the bunch as they receive it from the trader, fastening the ends at the back of the neck. They must be greatly inconvenienced at such times by the weight of their ornaments, for I have seen the young

women with such a mass of differently coloured
beads round their necks, as to occupy the whole
space from the bosom to the chin, and quite pre-
venting them from turning their heads; they wear
a petticoat reaching below the knees, made of either
their own bark-cloth, or gaily-hued printed cottons,
obtained from the traders; these are wrapped round
the loins, and tucked in on one side above the hip.
When "dressed" for company, they make the upper
part of the body a deep vermilion, a colour extracted
from the pod of an arnatto shrub; it is found between
the seeds, and when required is taken out and
collected in a little calabash, to be ready for use.
When rubbed into the skin it imparts to it a
soft and glossy look. The females do not paint the
face in broad bands of black and red streaks and
blotches like the men; but have, instead, three or
four very fine lines drawn evenly across the nose
and cheeks. In spite of the seemingly endless
variety of design in vogue with the men, Temple
assured me that they each have a recognized
meaning. I saw a Woolwa at Kissalala who had
his hair arranged in a very curious fashion; it was
tied up behind much in the same way as the old
European queue, but this was the sole example of
such a mode that attracted my notice.

On the present occasion, the Indians drank
mishla all that day and the next, according to their
custom, that they might leave none. During the
drinking, one of the party went round the circle
from time to time, singing a sort of monotonous

chant, beating a drum, formed from one of the joints of a large bamboo, to the accompanying notes of a flute of bamboo. Young Freshwater once, in making his tour, stopped in front of me, and I could not help smiling at his dolorous expression; but he stood still, and looked at me with unmoved gravity. This melancholy chant, or tune, seems to be the only Indian music, and in my subsequent journeys I often heard it whistled or hummed, when the Indians lay down for the night by the fireside, wrapped in their bark sheeting.

CHAPTER III.

ON the morning of the 14th I started for the Rama branch (which is an easy day's journey from Kissalala) in a canoe, with a young Woolwa from an upper settlement, Melville Temple, and a Woolwa boy. We dropped slowly down the river, and had to camp for the night on a very wet bank among the reeds, there being no better place to be found in the neighbourhood; fortunately, the night was fine. We rigged my mackintosh sheet (a most useful thing for travellers as long as it lasts, especially in wet climates) as a tent, and spread the damp ground with a layer of reeds cut with the machete. After enjoying a good supper and a smoke, we lay down under it; but, having encamped in the very haunts of the frogs, we were unable to sleep for the first hour or two, as they seemed to express, by a greater degree of noise than usual, their indignation at being thus intruded upon. The Indians laughed heartily when I called out to them as loudly as I could to "shut up," and the reptiles themselves seemed very much astonished at the unwonted order, for they ceased their drumming and shrieking, and there was a dead silence for some minutes. During

o

this lull the Woolwa said they could hear a jaguar on the opposite hill, but, though I listened attentively, I was unable to distinguish any sound.

We entered the Rama mouth early next morning: the view was much prettier than that in the mouth of the Rusewass, which we had passed the day before, and would have been beautiful had not all the trees been broken by the hurricane. The forest foliage in this part of Tropical America has a wild grandeur, very different from the gentle loveliness of our own northern woodlands, but I am not sure that the comparison is altogether in favour of the former: much of what is there gained in variety and immensity of its mighty trees, is rather marred in symmetry by the wild, matted tangle of flowering vines, and by the multitude of other parasites, which blend the whole into one gorgeous mass of flowers and leaves; whereas our oaks, elms, and beeches stand out in individual completeness and beauty of form.

We paddled on slowly up the river, shooting birds and iguana, which latter were more numerous here than on any other stream I know of; the overhanging trees, and the bamboo with which the lower parts of the river were clothed, were covered with them: they continually dropped from the branches into the water, just in front of the canoe, with a loud splash, like a shower of heavy fruit shaken off by the wind. The principal birds were several kinds of herons and kingfishers. We continued on until late in the afternoon without finding

a good camping place, the banks being everywhere
low, and covered with bamboo thickets thrown into
great confusion by the hurricane. A short distance
up the Rama there is a very remarkable conical hill,
standing rather far back, called by the Woolwa,
Assan-uka.

At length we arrived at a place where some
Indians had slept the night before, and my Indians
soon repaired their *waha* shelter. The waha is a
large leaf, not unlike that of the banana in shape,
but much tougher in texture; the tree grows in
damp ground, by the banks of rivers. The Woolwa
are very ingenious in hitching the leaves together
by their own split stalks to supports of bamboo, or
other poles, often those used to pole up the shallows,
thus forming in a few minutes a shelter that will
endure throughout a night's heavy rain or dew.

The next day being Sunday, I took a leisurely
breakfast, and did not start away till late. We
paddled along during the afternoon, Melville and
the Indians shooting iguana with their arrows for
our evening meal. That evening we camped comfortably on a flat rock at the foot of the first rapids.
A camp fire should never here be built on a rock
without first covering the bare surface with mould,
or there will be danger of explosion as soon as the
rock beneath becomes heated. Once, having neglected
to take this precaution, I was startled, as I lay on
my blanket at no great distance, by a sudden loud
report; the fire, together with fragments of rock,
was scattered about in all directions, and a monkey,

as well as some plantain which was roasting for dinner, was sent flying off into the river.

During our evening meal we scented an alligator, the proximity of which is always known by a strong odour of musk (like some European belles); and knowing their partiality for dogs, of which the native brutes are well aware, I was obliged to keep Jack close beside me. I remember an instance of the terror the canine race have for alligators: a dog belonging to Temple had a great dislike to the water, and one day, when he threw it into the river, it alighted near a dry rock; on this it immediately scrambled, refusing to return to the land, and it was only by practising on the animal's fears that his master succeeded in bringing him back. Taking advantage of a minute when the dog was looking in another direction, Temple hurled a piece of turf into the water, and the delinquent, thinking it was one of his enemies in pursuit, dashed into the water, and swam ashore at his utmost speed.

This part of the river is not without a sort of gloomy beauty; bamboo had become less common, and the banks, from which hung in heavy masses the foliage of gigantic swamp-woods, rose to a much greater height. These swamp-wood trees are a favourite resort of large iguana lizards, which lie on the thick boughs. That afternoon, when the men were trying to shake one into the canoe, which had been transfixed by an arrow, we were rather startled to perceive that a large and very venomous snake was coiled on the same bough. An energetic stroke

with the paddles, however, when we were aware of
the danger, carried us out of harm's way.

Next day we proceeded some distance further,
and passed an abrupt cliff of rock on the left bank,
rising perpendicularly from the water, which appeared very deep at its base. The current soon
became very strong, and as we were in a very
heavy old canoe, I determined to halt. Near this
point we fell in with a number of whistling ducks
(*Anas autumnalis*), seated in a row on a half-sub-
merged tree. Having shot one of them, the whole
flock flew round and round the wounded bird, until
we had secured five more. On turning back to
commence our homeward journey, we met two pit-
pans occupied by women who were going to the
first settlement on this river, which is a little higher
up. They had been fetching provisions from one of
their plantations in the vicinity. I stopped them
and bartered some birds for cassada, after which
we continued our downward course. The Indian
settlements here are few in number, and very high
up. The lower part does not suit the Woolwa, who
love to rear their houses on the elevated banks
above the river, although I believe they used to live
much lower down. That night we stopped at one
of our old camping places, and the next day continued our course under a broiling sun, there being
no shade to rest beneath, as the banks were very
low, with nothing but bamboo thickets on either
side. By the time evening came on we had left the
Rama, and arrived at a small rocky island in the

middle of the main river, called by the Woolwa
Assan-darkna, where we camped. The next morning we were soon on the water, and reached Kissalala
early. I was very glad to get back to my old lodge,
as during the last day, though I paddled along
resolutely, I felt decidedly feverish; indeed, I believe
that nothing prevented me from becoming worse
but the large quantity of sugar-cane I used. The
morning after my return my headache was gone,
but it left me very weak. This drawback, however,
did not deter me from going out on the river as
usual.

I now made an arrangement with Hercules
Temple to go with me into the interior, my purpose
then being to ascend as far as possible up this river,
which everybody, both here and on the coast, declared would bring me into the Spanish savannahs;
then to strike north, skirting the Spanish settlements,
until I reached the head-waters of the Wauks or
Patook, descend one of these rivers to the coast, and
so back to Blewfields, at which place I calculated
to arrive by the end of the dry season. I had provided myself before leaving England with a good
assortment of beads, fish-hooks, knives, &c., to give
to the Indians in exchange for provisions and other
requisites. Had I been able to carry out my
original intention, I should probably have fallen in,
more to the north, with tribes that are little known
(except by name) even to those longest resident on
the coast; as it was, my journey was confined merely
to the country of the Woolwa and Moskito.

Old Temple himself possessed something of the spirit of a traveller, and he spoke with considerable contempt of the Blewfields people, who knew nothing of the country beyond their immediate neighbourhood. As a young man he had traded much among the Blanco and Terribee Indians in eastern Costa Rica, and he related many interesting things about their mountainous territory, and the manners and customs of tribes whose names I had scarcely heard before. From his description I should imagine their manners and government to have been much like those of the warlike San Blas, on the Isthmus of Panama, but what their condition is at the present time I am unable to state. Like the San Blas, every village was governed by the oldest man in it, and their games seem to have been of the most martial and athletic kind, the umpires rewarding prizes to the successful competitors. His description of the San Blas was very interesting: their strange habits and hostility to the Spaniards, and their laws enforced to prevent foreigners from settling amongst them, although they are friendly with the traders, and freely barter the produce of their lands for supplies of different articles of commerce to the exclusion of spirits, which they do not touch. Many small vessels call at their coast yearly, especially from North America, to load with cocoa-nuts, grown by these Indians plentifully; and when the master of a boat wishes to trade, he is permitted to transact his business on shore, provided he joins his ship before night. Some of the men speak English, and

one, belonging to a party of San Blas I afterwards
met at Colon selling shells and parroquets to the
Spaniards, had visited the United States. They
were not unlike the Woolwa in person, being rather
below the middle height, strongly made, and thick-
set. There was much independence in their bearing
as they stood in their canoes at the water's edge,
appearing to regard their customers with supreme
indifference, whether they bought or not, confident
of getting sufficient purchasers for their stock before
returning to their fastnesses beyond Porto Bello.

I made an agreement with Temple that he
should go home to Blewfields to spend Christmas
with his family, taking with him the case of skins
I wished to send by the January mail; after which
he was to return immediately to Kissalala, when we
would, without delay, start for the interior. In
the mean time, I intended to keep myself quiet until
I had lost the slight feverishness and weakness with
which I had been troubled ever since my return
from the Rama. This used to come on about every
third evening, commencing with a chill and a
shivering, just as the sun dipped. I would make
up the fire and sit over it, and, being thirsty without
hunger, would brew a quantity of strong tea,
drinking it as hot as possible; this induced perspiration, and after spending rather a restless night
wrapped in my blanket, I awoke in the morning pretty
well, with the exception of a degree of lassitude.
I always consumed a great deal of sugar-cane after
these attacks, that being the only thing for which

I had any appetite, as I had taken a distaste to
the ripe plantains, and the different kinds of banana
that were procurable in abundance, being the only
fruit cultivated by the Woolwa, except shaddocks,
and a few large, but watery, oranges. They
usually eat one or two bananas the first thing after
rising in the morning. Cassada, if largely consumed, is said frequently to occasion heartburn.

This Christmas week brought my first sickness
since my arrival in this country. Neither an
approach to it, nor even loss of appetite, had
assailed me before. On Christmas Day I lay on my
lounge of split bamboo, not feeling at all well, and
watched the wood-ants driving their covered ways
along the supporting beams just above my head, in
the direction of a large nest they had established
in a corner of the hut. When these tunnels were
broken, the inhabitants immediately swarmed to
the breach and commenced to repair them. I could
hardly realize that this hot, bright day, with the
vegetation greener than I had ever before seen
it, was indeed Christmas Day; and of course,
in my loneliness in the foreign land, thoughts of
home, and the friends from whom I was so far sepa-
rated, thronged upon me. The next few days,
being much better, I began to grow very tired
of this forced inactivity, and to look forward
impatiently to Temple's return, that I might proceed on my journey. He was longer absent than
I had expected, and I was obliged to while away
the time as best I could, between fishing, shooting,

and observing the mental and physical peculiarities
of the Indians.   Though living in such a retired
spot, these people are very particular in the require-
ments of certain forms of etiquette among themselves, as I discovered before I had been long
in their society.   One day, soon after my arrival at
Kissalala, I was taking a constitutional turn round
the lodges, while the men were away in the canoes
fishing.   Seeing a woman who had often brought me
plantain, banana, cassada, &c., I nodded, and
wished her " Good morning."   I shall never forget
the scared and astonished look that appeared in her
face, and at once I comprehended that I had been
guilty of a falling off from good manners.   After
the first surprise, the dusky lady seemed to recover
her usual presence of mind, remembering probably
that I was but a stranger from some distant land of
barbarism, and therefore unaccustomed to polite
society.   Accordingly, she recommenced busying
herself at her interrupted domestic duties, while
I made the best of my way down a steep bank, rod
in hand, to capture a fish for dinner, in my favourite
spot.   This was a shady nook, beneath the thick
foliage and long-twisted limbs of aged swamp-
wood, which grew near the landing-place where the
Indian canoes were generally moored.   A little
streamlet, rising somewhere far back in the depths
of the forest, here broke from the deep gully that
formed its channel, and fell with a gentle murmur
over a rocky slab into the river.   The place was
delightfully cool even on the hottest day, and by

dropping my hook, baited with a green grass-hopper, just where the water rippled in silvery bubbles beneath the little cascades, I seldom failed to secure a fine "*shirrick.*" This was also a favourite resort of birds: many species of yellow fly-catchers perched upon the topmost twigs of the swamp-woods; humming-birds darted round their thick boughs in the time of blossoms; and brilliant crimson and black finches fluttered in pursuit of one another in the more shady thickets. The swamp-wood trees on the river's bank, when blossoming, are havens for many varieties of humming-birds, which almost dazzle the eye with their rapid flight in and out among the heavily foliaged boughs—now poised on swiftly vibrating wings over the clustered flowers in which they secure their minute insect prey, and again darting away with a velocity that baffles the powers of vision; or perching composedly upon the topmost twig, to plume their ruffled feathers. The tiny creatures are very fond of frisking together and engaging in mock or real fights. When they dart close by you, the clear hum, from which this peculiarly interesting family of tropical birds derives its name, is very perceptible.

A fine crested eagle is not uncommon here, and a great variety of yellow-breasted fly-catchers were to be seen everywhere on the river's bank. They are called "*kisscadee*" by the Indians, with whom they appear to be great favourites, probably on account of their pretty forms, and gentle, though fearless disposition. A little martin *(Cotyle wropygiatis)* frequently perched upon the bare top of a broken tree near my lodge.

I was surprised to find that the natives of this country do not know how to manufacture tobacco. The coarse leaf-tobacco used by them is imported from the United States.

CHAPTER IV.

TEMPLE returned on the evening of the 22nd of January, and I was truly glad to see him, having spent a weary time waiting for him. He had been detained by cholera having shown itself in some members of his family. By his account, this disease, which subsequently wrought such fearful havoc in some of the Moskito villages, on the coast more to the north, had been brought to Blewfields. Cholera has often visited Grey Town; and in December, 1866, it broke out with such fury on board one of the American river steamers belonging to the Transit Company, on the San Juan river, that it had to lie by in one of the creeks, and was soon deserted by the remnant of the crew and passengers. Some half-breed Moskito men who were at Grey Town heard of this disaster, and started up the river to plunder the vessel, from which they took some provisions, clothing, and other things, returning toward their village in the north with their booty; but before they reached Monkey Point, one of their number was seized with the disease, died, and was thrown overboard. On arriving opposite Blewfields Lagoon, another died, and they put into the bluff to bury him, which was done in such haste that the legs of the body were left exposed to the sun; then they continued their journey to the north, no doubt sowing there the seeds of the harvest which resulted so fatally. I was told of one village in particular where a great part of the women were spared. A lad, having paddled over to the bluff from Blewfields to cut cane, was attracted to the spot where the man had been buried by the stench, which he probably thought proceeded from a dead cow; but approaching the spot, he saw the man's legs above the ground, which gave him such a shock, that he returned at once to make known the discovery. As it was Christmas week, he went to a dance in the evening, the custom of these people being to go in a party from house to house, until they have danced in all the houses of the village that are large enough. While still at one of these houses he was taken ill, and died before morning. The disease attacked several other persons in a greater or less degree, but there were not many deaths, possibly on account of the cleanliness of the village of Blewfields, and the distance of the houses from each other. This catastrophe accounted for the length of time Temple had kept me waiting, as, naturally, he did not wish to leave his family while they were in danger.

When Temple went to Blewfields, he took with him an Indian lad from Kissalala, to assist in paddling his canoe. On the news reaching us that the cholera had broken out at Blewfields, the two sisters of the young man immediately took a small pit-pan, and started to fetch him home, out of danger of the dreaded "sickness," as the Indians call this terrible scourge, so different from their known diseases. When the girls arrived at Blewfields, they found Temple about to return with their brother, so they all came up the river together, without any harm apparently accruing from their visit; the two women cooking supper when they camped, and paddling their own little pit-pan. After their return, one of them was taken sick in the afternoon, and got worse during the night. An Indian came to me for some medicine, and I gave him some essence of ginger in a little hot tea (the only specific I had with me). This seemed to relieve her, but soon after they gave her some mishla. She again became worse, and just before dawn, while lying awake on my bamboo couch, I heard the crying of the women, by which I knew that she was dead. The Indians clustered into Temple's place, appearing very much startled, and all the fires began to throw out heavy wreaths of smoke, from the fuel of a bush which is burnt green, as a disinfectant. They now seemed to think that "the sickness" was fairly among them, and some guests, who had come from the very head of the river, instantly hurried their things into the canoes, and I heard the rattle of their paddles while it was yet dark. At daybreak the other sister, who had been to Blewfields, began retching violently, and, creeping down the steep bank to the water's edge with great difficulty, with the aid of a staff, died there in about two hours. The people now became so alarmed, that none but the old mother of the two girls and their brother would approach the bodies. All were seized with a regular panic, and came to me for remedies. Many of the women were retching, and Teribio, the Moskito man, looked quite pale, and complained that he felt very sick. My chattels were fortunately packed ready for my journey, so Temple and I encouraged them in their wish to leave the place immediately. I gave them a good dose of essence of ginger all round, which, if it did nothing else, partly reassured them. Having cautioned them against the *mishla*, they capsized a whole row of brimming pots that had been prepared for the morrow. Freshwater and Jackson had already paddled off with their families, and the remainder got into their canoes and followed, Temple and I occupying one together. I was much disgusted with Oosi-Maria and another Indian, the husbands of the two women who had died, for after cutting off their hair in sign of mourning, according to the Woolwa custom, they were so panic-stricken, that they joined the rest of the fugitives, leaving the mother and the lad alone with the dead sisters in the deserted village. This heartless behaviour must have been the fruit of their unconquerable fear of the strange "sickness," as the Indians usually have a funeral ceremonial, and much mishla-drinking. There is a regular place of burial, and at funerals a long line of spun cotton is stretched, like a telegraph-wire, from the house of the deceased, where the mishla-drinking is going on, to the interment-ground (where the body has been deposited), no matter how distant it may be. I have seen it following the course of the river for many miles, crossing and re-crossing the stream several times.

Captain Lewis, in his interesting book on the Wild Hill Races of South-Eastern India, mentions a similar usage among the Tipperalis of India.

The whole community of Kissalala camped that evening on a rock below the Matuck rapids, and next day we paddled on till the afternoon, passing many rapids, and often being obliged to haul the canoe. On camping, we made a very good meal off a *warry* (wild-hog), which Teribio had killed. I did all I could to rouse the Indians from their terror, but the exertions of the day seemed to have the most beneficial effect. They always used a quantity of red *chiti* (pepper), · which is here common, in cooking; but since their fright they had made their soups and boiled meats, usually very good, so hot, that I could hardly touch them.

We passed a fall without much difficulty, and, the next day being Sunday, remained in camp. Freshwater killed a most beautifully-marked *ocelot*, which we afterwards found lying on the rocks below. It had been too long exposed to the sun, so I could not secure its skin, which was a great pity, as I have never seen another specimen of equal beauty.

It was amusing to watch Teribio's little daughter

P

towing one of the pit-pans in the shallow water, and kicking about like a young otter. She was seven years old, and betrothed to young Freshwater, in conformity to the Woolwa custom of the selection of wives when mere children. The future bridegroom resides with the father-in-law elect, and superintends the education of the young lady. This little bride in prospective was an only child, a great pet, and, as is usual in such cases, very much spoilt.

28TH.—I manned a pit-pan with the men I had engaged at Kissalala, and continued my journey. Besides Temple and myself, there were three men in the canoe, viz., Teribio, the Moskito man, and the two Woolwa, Oosi-Maria and Jackson; the rest of the people we left in the camp. They had arranged that Freshwater and another man should take Teribio and Jackson's wives, with some other women and children, up a little creek, that fell into the river just above the beach on which they were camped, to remain there out of the way of the "sickness," until Teribio and the two returned, after we had reached the head of the river.

We passed the night in an Indian encampment, which we discovered pitched on a gravel bank, among some rocks, in the midst of the river, now become little better than a succession of rapids and falls. These were the first people we had encountered since leaving Kissalala, as at all the other places we passed the Indians had fled far up the little creeks at news of the "sickness," generally

leaving at the mouth of the creek a wand, with
a piece of white rag fluttering at the end, to indicate the direction they had taken.   On the following day we passed the Moroding Falls, up to which
point sharks are said to ascend the river.   We found
that the pit-pan we were in was too heavily laden
for this rapid and dangerous part of the river, where
we sometimes had to make several portages in a
day.   I authorized Temple to bargain with these
Indians for another, and also to engage two more
men.   In this second pit-pan we carried the heavy
bunches of plantain and other provisions, besides
some things which we removed from the over-laden
canoe.   I remained in the larger one with Teribio,
Oosi-Maria, and Jackson, while Temple took the
command of the other.   We were kept in this place
till the next day, making these arrangements and
collecting provisions.

I always knew when the Indians alluded to me
in their conversation as we travelled along : they
invariably spoke of me as *Waikna*, literally "man";
and as I lounged upon the seat in the middle of the
pit-pan, engaged in taking notes, or skinning a bird
during our morning paddle, it appeared to me that
I often afforded subject-matter for long and knotty
discussions among them.   Soon after starting on
the morning of the 30th, we arrived at a place
where the river is quite blocked up, and lost to
sight, amidst great stones fallen from the side of a
rocky hill.   Here we had to convey pit-pan and
everything over the hill, by a long and steep, but

well-worn portage-path through the forest, striking
the river again on the other side, as it emerged
from the obstruction in a rocky gorge : the banks
were densely covered with vegetation, resembling,
in this respect, every other part of the river's
course. After one more portage we came to an
Indian settlement, where they gave us some cassada;
and, bidding them adieu, we camped a little further
on. When passing through the channels among
the larger boulders in the river's bed, we often disturbed flights of small bats. The effect was very
strange, as they would flit like arrows shot from
an unseen hand, for a short distance, to the shady
side of some rock, into which the flooding waters
had worn curious cavities : here they seemed as
quickly to vanish as they appeared, for when they
had once fastened themselves to the crevices, their
colour and shape so much resembled the unequal
surface of the cross-grained stone, that it was almost
impossible to distinguish them even at a limited range.

During the next morning's paddle, we shot three
monkeys and two *quam* (or *guan*), a bird about
half-way between a fowl and a turkey in size; so
we stopped and had a good dinner under some
shady swamp-woods. That day we passed beyond
the site of the late hurricane. Since leaving the
First Hill, as the long portage we passed the day
before is called by the natives, its effects had not
been very noticeable. There was a very great
change for the better in the appearance of the
woods. Many noble silk-cotton and other trees

shoot out at the numerous bends of the river in their true and unmutilated proportions, throwing their giant limbs far over the stream. The tall, tough eboe is not seen much further from the coast than where the influence of the tide is felt, that is to say, not much beyond Kissalala on this river. The eboe-nut is a great favourite with the gorgeous macaws, which come from a long distance to feed upon it, and it is also useful to the people on the coast, who make from it a very fine oil. The bamboo now began to be seen more plentifully, increasing in size as we ascended the stream. The loud note of the "partridge," as the creoles call it, was to be heard in the dense forest on either bank; but I was never able to catch a glimpse of the bird, although two of the eggs were given me: they were blue, like those of the hedge-sparrow. There are two kinds of curassow which inhabit this region—one, called by the natives the queen-curassow, beautifully checked all over, like some of the bitterns of the country, with deep brown and black markings upon a ground of light warm brown; the other is black and white: both kinds have the head ornamented with a handsome crest of curled feathers, which the Indians often convert into a tasteful head-dress.

On the 1st of February we camped at an Indian settlement. In the dry season, when the water is low, the Indians at all the settlements encamp on the rocks at the water's edge. They live in their regular substantially-built houses on the high banks

above only in the rains, when the river is subject
to great and sudden rises. The next day we passed
an exceedingly difficult part of the river, and
came to the Second Hill, where the stream was
entirely lost amid the immense rocks. The portage-
path was not so long as that at the First Hill,
but equally well-worn, and a very pleasant one,
leading through the shady woods. It is noticeable
that where a portage has been used, the cavities
in the rocks are filled with cedar-shavings, some old
and some fresh, scraped off the bottoms of the pit-
pans by the rough surface of the stones. On
arriving at the river at the other side of the hill,
we cooked our dinner on the gravel bank at the
mouth of a shady little creek, called Billwass, which
here joins it. I was told that Woolwa lived up
this secluded little stream; and on more than one
occasion I have observed their predilection for these
out-of-the-way creeks, which have obstructions in
their mouths even for pit-pan navigation. As soon
as we had dined, and the Indians had refreshed
themselves after their exertions, as is their wont,
by bathing in the rapid water, we again pushed
on, finding the river still very difficult, on account
of the rocks with which its course was impeded.
Before nightfall we arrived at a settlement which,
I afterwards learned, goes by the name of Woukee,
as it is situated at the foot of the falls of that name
(so called after a kind of bittern), part of which can
be seen from the lodges.

The next day was Sunday, and I remained at

this place to give a rest to the men, who, for the last two days, had had very hard work with the poles, making portages, and dragging the canoes over the rocks. I am sure if some of those who condemn Indians as a lazy race had seen them at this work they would have revoked their judgment! Woukee is decidedly the prettiest settlement on the river, from the manner in which the houses are built, and the grounds planted around them. Besides the universal *Supa* palm, which, when on the banks among the forest trees, always indicates the sites of the old settlements, and others, usually seen among the Woolwa, a fine bread-fruit, various fruit trees, and also a large quantity of chocolate and cotton, are planted here. The old man who seemed to be the patriarch of the place, was lame in one leg, from the effect of a snake-bite, from which he was just recovering. I had hitherto considered the men of this tribe as living in the seclusion of their rivers and forests, but this old man proved to be an exception : he had travelled to a considerable distance, having in his youth engaged himself to traders on the coast as far south as Salt Creek, in Costa Rica, and among the San Blas Indians. Temple recollected seeing him with the Teribee and Blancos in Costa Rica : one of the men in the provision canoe had also travelled a long way, having been through the Spanish territory about lake Nicaragua, to Granada, Leon, Massaya, Managua, and other places. Many of the Woolwa at the head of the river understand

Spanish, whilst most of those of the lower river,
who go to Blewfields, speak Moskito ; but this man
was conversant with both languages. About the
centre there were, however, many Woolwa who
knew no tongue save their own.

During the day I spent here, all the Woukee
people were busily engaged in grinding the cane
and making sugar, the young men turning the stiff
rollers by hand-spikes, while the patriarch, assisted
by some girls, placed the cane between the rollers.
The juice, on being thus crushed out, ran down
banana leaves into one of their large earthenware
pots.

Temple and I remained at Woukee till the next
morning, when we joined Teribio and his companions, who had already taken the pit-pans and
their loads over the succession of rapids and falls
above the settlement. We had to walk through the
skirts of the forest along a rugged track, with sharp
pieces of rock spiking out of the ground everywhere,
and covered with fallen trees; then by the water's
edge, climbing from one mass of water-worn rock to
another, until at last we reached the canoe. I found
this kind of walking very tiring to the feet, for in
these canoe-journeys I followed the Indian fashion,
and went barefoot, as it was quite the best plan,
in consequence of the quantity of water that came
over the sides and flooded the bottom of the canoe
whilst we navigated the rapids. When we were on
the point of turning a corner, I heard loud cries in
the rear, which I took to be those of monkeys in the

adjoining forest; but the men understood that the
Indians, now in the distance, were hailing us to
return; on putting back, we discovered that in
scrambling over the stones I had dropped my note-
book and map; these they had picked up, and came
after us, at considerable inconvenience to themselves,
to restore them.   This is but one of many instances
of the scrupulous honesty of this tribe of Indians;
indeed, to the best of my knowledge, I never even
lost a string of beads or a fish-hook whilst living
among them, although they had abundant oppor-
tunities for theft when I was at Kissalala and
other places, for I had many things lying about
that must have seemed extremely valuable in their
eyes.

At a mid-day halt once, we happened to have
nothing to eat in the shape of meat.   Temple consequently fairly lost his temper, and annoyed me
considerably by grumbling to the Indians; so I took
him aside, and told him if he was not contented to
put up with what I did, I should send him back to
Blewfields when we arrived at the next settlement.
I heard no more of his murmurings, but was not
sorry when, after we had gone a little farther, we
secured a couple of *quam* for dinner.   Animal life,
contrary to my expectations, was scarce on this
river, rather decreasing the higher we ascended.
Now and then we startled a flock of shag and darters,
balancing themselves clumsily on the bare limbs of
a tree over the water. Sometimes our approach
would so alarm these stupid-looking birds, that

they altogether lost their equilibrium, and fell with
a loud flop into the water, where they made the best
of their retreat by a long dive; occasionally a
solitary white heron, a teal, or a Muscovy duck,
rose from under the bank; or a large black crab-
catching hawk broke the stillness with his peculiar
whistling note. At night, too, we sometimes caught
a pair of brilliant little trogons, sitting motionless
under the shade of a swamp-wood. Snakes were
common on the banks, some of them beautifully
marked, such as the *barber's pole*, for instance;
but iguana, so plentiful lower down the river, had
become scarce. At times the men would whisper
that they saw deer on the bank ahead, but they
always disappeared behind before we could get
nearer. The tracks of the tapir, or *mountain-
cow*, were very numerous on the banks, showing
where they had crossed in the night, for their habits
are nocturnal. That night, after we had lain down
to rest, I heard the cry of an owl, which at first I
imagined to be some one calling to us from the
woods, so exactly did it resemble the human voice,
hailing us in most unfriendly tones from the depths
of the gloomy forest that shut in the river on each
side. On the upper part of the Blewfields river, a
large species of owl *(Syrnium perspicillatum)* was
often pointed out to me; they were detected by the
sharp eyes of the Indians, sitting in pairs, under the
shady plume-like clusters of the tall bamboo: some
of the older ones which I shot were armed with
most formidable talons. The little fin-foot is very

common on the Blewfields river. This bird is very similar in appearance and habits to our dab-chick, or small grebe; it is generally to be seen paddling about under the banks, or skulking among the reeds; its flight is heavy and of short duration, but it is an adept at diving: the creoles call it the *diving dupper*.

On the 5th, we hauled the canoe over an immense rock blocking up the passage, and on which we cooked a *karkee*, or Indian rabbit, shot in the morning by Blue-blossom, one of the men in Temple's canoe.

We stopped for the night at a considerable Indian settlement, the largest, I believe, on this river, consisting of five or six houses on each bank. From what I saw here as well as at the other places, I came to the conclusion that the Indians have many narrow paths through the forest known only to themselves, leading far away, and perhaps connecting this with other rivers. They are intimately acquainted with every little creek and watercourse of their country; and I think any who may be interested in the correct mapping out of the districts, would do well oftener to consult the aboriginal Indians, instead of the mongrel caribs and creoles on the coast and in the Spanish country. At this settlement I saw the first Woolwa with grey hair that I had encountered—a most pleasant-looking, contented, old man, who, I was informed, was the oldest man on the river.

I expected that the next day would be my last of

the water before starting across country. As, however, the morning threatened much rain, we remained stationary until the following day. Teribio killed a fine deer on the evening of our arrival, which, with some *quam*, supplied our larder in a very satisfactory manner; for the venison in this locality is exceedingly good. There is a sort of tick in this country, called by the Spaniards *garra-patas*, which infests the vegetation in the dry season. These insects are a source of great annoyance to every one travelling through the woods, as they are easily brushed off by the clothes, and immediately fix themselves on various parts of the body, burying their heads in the skin, causing great irritation. Spanish travellers in these regions often carry with them a lump of soft wax, with which they extract the " garra-patas," by pressing it on the spot where the insect is embedded. Dogs and horses suffer very much from them, and very large ones are often seen on cattle and deer; they sometimes get into the nostrils, and mount so high that it is not easy to reach them. All this night I was made very restless by a creeping sensation stealing all over me, like the crawl of some unfamiliar insect; in the morning I found myself covered with ticks. Temple was as surprised as I was at finding them in an Indian lodge, and inquired of Teribio the meaning of the invasion, who laughed, and pointed to the deer's skin suspended from the rafters over my sleeping place, which, of course, explained the phenomenon; for the deer, as a denizen of the forest, is always covered with them, and the insects had been dropping down upon me all the night. I was engaged for nearly two days in picking them out.

While we were at the village, I sent Teribio and Blue-blossom into the thick woods at the back of the houses, to shoot some of the birds for me that lodged in the recesses. They brought a few large and small trogons, and a bird which I had not seen on the river before *(Eurypyga major)*. During my stay here, I had a glimpse of what might be called a very ideal of savage beauty. I had already seen several strikingly handsome young men, but there was a grandeur about this youthful head, with its heavy masses of jet-black hair hanging over the forehead, and throwing a deep shadow over fine aquiline features, and large eyes of wonderful brilliancy. He surpassed anything that I had before seen of the natural nobility of man. When I turned to take a second look at this beautiful youth, he had disappeared from the lodge, and I saw no more of him.

On the 7th we arrived at a place called Kaka, which we heard was the last settlement of the Indian tribe on this river; but I doubt the fact; for though the river had become a small stream, there must still have been, for some distance further, much more water in it than in some of the creeks lower down which are inhabited.

In the afternoon a man entered the lodge in which we had made our fire, strumming some very pretty airs on a sort of guitar. He spoke to us in Spanish, and volunteered to give information about the country lying before us, offering himself as our guide, and saying that the nearest places in the Spanish country were Consuelo and Libertad, to both of which the Indians had tracks through the forest from this settlement, enabling them to carry plantain and banana for occasional sale. Hearing from him and the Woolwa that Consuelo was the nearest, and also that there were "mucho Inglez" there, I decided upon walking over to it with Temple, though I did not give much credence to their assurance of the number of my countrymen I should find, thinking that the Indians might imagine all the Europeans who could not speak the Spanish language to be English. But there might, perhaps, be one Englishman in the place sufficiently acquainted with the country to be able to give us some information about our position, and directions for our future course.

My men found some boon-companions at Kaka the evening of our arrival there, and in the congenial society drank themselves very noisy with the fermented juice of the sugar-cane.

Kaka is a pretty little place, lying embedded in woods; and on the opposite side of the river, here very narrow, rose an especially beautiful wall of verdure, the tall, straight shafts of the trunks and limbs of trees appearing at intervals between the masses of varied foliage,—flowering vines, caught up here and there in festoons, or hanging down in heavy tresses, the snake-like coil of the bush-ropes, and the elegant fronds of a palm, occasionally breaking the outline.

On one side the lodges ran into what would, in Wales, be called a trout stream—the same slabs of rock, deep pools, little falls, and brawling hollows; but here, the palm, the tree-fern, and many other peculiar forms of vegetation known only to the tropics, overhung the glancing water; while from the green walls that shut in the stream on either side, and through which only a stray beam of sunlight found its way, came the strange cries of parrots, toucans, trogons, and other birds, whose voices were in harmony with the scene.

I saw an exceptionally large sugar-cane plantation at Kaka; and the Indians make a very palatable kind of sugar, moulding it into cakes to eat with their plantain and baked cassada. I soon followed their example, and found it a great improvement to the fare. We had much rain during the latter part of our journey, but were not deterred from our intended progress. The day after our arrival was spent in preparing for our travels by land.

CHAPTER V.

On the 9th of February, Temple and I, with some
Indians, started for Consuelo directly after our
morning cup of tea, which I always took at daybreak whilst travelling. The track, in some places
very faint, led through a damp and gloomy forest,
several times crossing a creek (the same that joins
the river at Kaka), thence up the sides of a high
steep hill, covered with tall straight trees, closely
intermixed with others of a shorter and thicker
growth, saplings, and dwarf palms, the whole
bound together at intervals with bush-ropes. It
was impossible to obtain any view of the surrounding country until we arrived at the very summit,
where there was a small open plot of grass. Suddenly emerging from the tangled forest, rather
exhausted by the wearisome climb, in which we had
often been obliged to creep under, or scale the trees
and branches that lay across the track, we came
unexpectedly on a view of great extent and beauty:
the plain beneath, diversified by hills of different
elevation, stretched far away to the foot of distant
mountains. The day was so unusually clear, that

the Indians could point out on the distant slopes the savannahs, distinguishable by their light-brown tint; they called them the mountains and savannahs of Matagalpa. I was never again able to get a clear view from this eminence, as on subsequent visits the woody sides of the hill were enveloped in mist and vapour. On our left, through the trees, we caught a glimpse of a curiously-shaped rocky summit, a little higher than that on which we stood, and which I afterwards discovered to be the peak of Peña Blanca.

We only met with one snake in our passage through the woods, and that a small, insignificant brown creature, though the Indians appeared to hold it in great dread, as they took the trouble to beat it to death with a long stick, throwing it out of the path before they would pass the spot. I only came into actual contact with the reptiles on two other occasions, although I have often shot them on the bank from my canoe, at the request of the men. I killed a snake once on a path near the Javali mine in Chontales; and another time, at Kissalala, when sleeping by my fire, I felt something on my blanket at my feet; I kicked out drowsily, and then waking up, I saw by the fire-light a snake wriggling off as fast as he could, and ere I could fling a brand at him he was gone. Before this incident I had been under the impression that snakes were afraid to approach a fire. Of other living creatures also we saw very little in this journey.

After leaving the open space on the summit of

the hill, we did not take long to go down the steep
descent on the other side, which led to a narrow
valley, evidently but recently cleared of its timber.
I observed during our fatiguing march through the
forest the sun was always before us, by which we
knew that we had been going nearly due east since
leaving Kaka.

We paused to make inquiry at the first hut we
came to, after leaving the woods; it was inhabited
by Spaniards. I asked them if there were English
at Consuelo; whereupon one of them politely offered
to conduct me to the house of an Englishman. On
our walk thither, Temple and I saw enough to con-
vince us that we were in a mining settlement of
considerable importance. Having reached the house
we sought, our guide pointed to a man sitting in a
hammock, while two women of English exterior
were cooking at a stove what looked more like
beefsteak than anything I had seen for a long
time. Turning to the man, I asked him if he were
English. He replied, evidently surprised at the
question, "I should rather think so!"

Judge of my amazement on finding that I had
come out of the forest in the midst of the Chontales
mines, which from the map I had supposed to be
due south of the Blewfields river. It was, however,
equalled by the astonishment of the people here on
seeing us come out of the forest; they are so shut
in by densely-wooded hills that they never dreamt
of the possibility of any one arriving among them
except by the road from Lake Nicaragua, through

the town of Libertad. The hearty old Cornish
mining Captain took me to his room, where a dinner
of beefsteak and bread was already on the table,
of which I was very glad to partake, as my morning's
march over the steep hill had tired me much, after
accustoming myself only to canoe travelling for so
long.

The next day I went to San Domingo, the head-
quarters of the Chontales mines, where I was very
kindly received by the officers of the mining company, who seemed much surprised to find that they
were only divided by the hills behind the settlement from the large river that flows into Blewfields
Lagoon. At this place I saw an immense number
of the vultures called " John Crows " on the coast—
" *Zopitotas*" by the Spaniards. In the evening I
returned to Consuelo, where Mr. Longland, the
purveyor, placed a hammock at my disposal. On
the morrow I went back to Kaka, and paid off all
my Indian men, except Teribio and the two Woolwa
from Kissalala, who said they would rather wait till
I returned to Blewfields after my journey, and then
be paid "in cloth," as they call printed cottons,
since I had nothing with me that took their fancy.
I thought this showed considerable confidence in
me, as my word was their only guarantee that I
should return to Blewfields.

I went to Consuelo the next morning, leaving
Temple at Kaka in charge of my things, which I
left there until I could arrange to have them brought
over the hill. It amused me highly to see the

Woolwa who came with me take off their simple *toumoo*, and put on trousers and shirt, before coming within sight of the first house. I do not know their reason for doing so, as at Blewfields they constantly walk about with nothing on but the *toumoo*.

I dined with the late Captain Hill, R.N., then in command of the mines; he gave me some very interesting particulars of a cruise among the islands off the unfrequented coasts of New Guinea. Hearing that the man engaged to go with a pack-ox to Kaka for my things had not yet started, I went to see what delayed him. These mongrel Spaniards are the most tiresome people to deal with imaginable; if you attempt to hurry them, they reply, in the coolest manner possible, "Poco tiempo." When I was on the river, some troublesome little ulcers had shown themselves about the ankles, and had since become much worse; so the doctor at the mines advised me to rest until they were better. I was not able therefore to take many birds among the surrounding hills, or to see as much of the country as I desired. The officers' quarters were at this time in a wretched native building, composed of mud walls, boarding, and thatch. One evening, when I was sitting there as usual, a man came running in, much excited, saying that a person named Clato had been stabbed while gambling in the carpenters' workshop, the great resort of the natives on pay-night. Of course we all took our revolvers and hastened to the scene. The shed in which they had been gambling was so

thronged that we could hardly elbow our way up to
the work-bench on which the wounded man lay. I
recognised him at once, as a tall, gaunt Spaniard
whom I had seen in conversation with the native
miners as they came from receiving their pay—
probably arranging the evening's play which was
to cost him his life; for a single glance at his livid
face was sufficient to show that his minutes were
numbered; and so it proved, for before we left the
shed he was a corpse. When we entered, the
doctor was doing all he could for him; but though
his wound between the ribs looked wonderfully
small, and there was very little blood to be seen,
the internal hæmorrhage must have been great, for
he was very soon choked. I could not help thinking,
whilst looking on the powerful frame before me,
laid so low, and by so small a thing (a pocket
clasp-knife, afterwards found in the shavings in
the shop), how easily the "silver cord" is loosened.
It seemed that none had witnessed the fatal
blow, though the Cornish Captain, on hearing the
disturbance, had gone in with his revolver to
disperse the disputants in time to see the Spaniard
fall. The next day, an officer, with some Nica-
raguan soldiers, arrived, and made inquiries into
the murder; in consequence of which about a dozen
men, witnesses and petty offenders, were put
into the stocks. They did not secure the murderer,
who, of course, had made his escape into the
bush.

On the 20th, I was informed that my messengers

had returned from Kaka without bringing my baggage; on inquiry, it turned out that Temple, for some unexplained reason, had left the place, and the Woolwa would not give up what was in their charge to strangers. The honesty of these aborigines contrasts very favourably with the thievishness of the Spaniards, who stripped me of nearly all my little nicknacks during the short time I remained at the mines. I was much annoyed with Temple, having given him orders to wait at Kaka until I either went myself or sent some one else in my stead. I engaged one of the Matagalpa miners to accompany me, and set out for Kaka the next day; but when, after wading through the creek, we entered the place, it was quite empty. I fired off several shots from my revolver, and then, nobody appearing, I desired the Indian to take my axe, chop some wood, and cook some of the green plantains that were hanging up under the eaves of the lodge in which my things were deposited. When this was done, and I had prepared some tea, I invited him by a motion to join me, and we sat down to enjoy our frugal supper. I made some attempt at conversation with my companion in broken Spanish, but he did not seem to be very quick at understanding. At dusk, a strange Woolwa appeared above the bank, coming from the river, attracted no doubt by the light of a fire at a place that he probably knew to be empty; he was in full dress (paint, feathers, &c.) and very tipsy, which at once explained the absence of the inhabitants at a feast in the neighbourhood.

I was not sorry when he staggered down the bank to the canoe in which his wife awaited him, for his good-tempered familiarity was not very pleasant. The next morning the missing inhabitants trooped in, and I engaged two young men to carry my two tin boxes over to Consuelo.

I was much amused by the assertion of my Matagalpa friend, who, when asked by the Woolwa what nation he belonged to, declared, with great emphasis, that he was an " Espagñol claro," though he was, in reality, as much an Indian as any of his Woolwa questioners. I was struck by the facility with which they made themselves understood, by rapidly exchanged signs, although totally unacquainted with each other's language.

On our way back to Consuelo we met Temple, and several Woolwa boys returning to Kaka; his surprise was great that I should have walked so far with my ankles in such a painful state. To excuse himself for his absence, he said that he had thought it better to employ the time, while waiting at Kaka, in making inquiries about mules for our journey, and, for that purpose, had followed the other track to the town of Libertad; but he had been unable to return sooner, having been attacked by a violent fever. I told him, however, that his good intentions were no excuse for leaving, without an order from me, the things I had placed under his care.

During the remainder of my stay at the mines, I messed with the officers at their quarters, and

formed some sincere friendships; and I still entertain the hope that I may one day have the pleasure
of again meeting Mr. Gill and the kind Californian
lady, his wife. In the latter part of February and
the beginning of March, the wounds on my ankles
became very painful, owing, perhaps, to the rough
treatment they had received on my last walk to
Kaka, and being often bruised against logs and
stumps; but I was assured that nearly all the English
in the mines had suffered more or less from these
troublesome little ulcers. I now began to despair
of carrying out my original intention of a northerly
route, as the dry season was very far advanced. If
I had waited for the heavy rains, commencing about
the end of May or the beginning of June, all my
own things, as well as any collection of birds that I
might make, would have become quite spoiled before
I could have brought them down to Blewfields.
Even now, so great was the dampness of the climate,
I had much difficulty in preserving those I had
already obtained. Some water-colour paper for
sketching was so damaged that it was impossible to
use it, the colour running as it would on blotting-
paper; so that I was obliged to take my sketches
in pen and ink.

Captain Pim, R.N., arrived at Chontales for
a visit whilst I was resting there. He appeared
to be much interested in my account of Blewfields
river, and seemed to entertain, with Dr. Seeman,
of the Javali mine, the idea of opening communication with Blewfields by means of a mule track, cut

through the forest to Kissalala; from which place
they would have a small craft down the river, the
lower part of its course being free from obstruction.

This plan would be a great saving in point of
distance, for at present the only communication
with the coast is by way of Lake Nicaragua, the
San Juan river, and Grey Town. If the scheme
should be carried out, the Nicaraguans would probably make an effort to gain possession of Blewfields, as they did formerly in the case of Grey
Town. It is to be hoped that the Government of
England would deal more honourably by our old
Moskito friends than it did in the treaty of 1861,
of which, whenever it was mentioned (and the
natives frequently questioned me about it), I felt
very much ashamed.

The weather was now delightful, and as my
ankles were a little better, I determined soon to
bring my stay at Chontales to a close; and as it
was not possible to go round by the Patook route
this season, I resolved to return to the Moskito
shore. Temple assured me that the Blewfields and
the Pearl Cay Lagoons, as well as the neighbouring
creeks and rivers, at this time of year abounded in
various kinds of water-fowl; and I purposed making
there the best collection of birds that I could,
during the remaining dry months, April and
May.

The forests about the Chontales district look
very heavy and gloomy, and walking in their

shadows causes an oppression on the spirits very
different from the lightness of heart with which
one treads the sward beneath the "greenwood trees
of merrie England." There is also a singular
scarcity of animals of the larger species. Deer,
wild hog, even curassow and *quam*, are by no
means plentiful. From the state of my ankles, I
was not able to penetrate far into the woods in my
search for birds; but I sent Temple occasionally.
I noticed two or three varieties of toucan, one of
which was larger than the *Ramphastos piscivorus*
I saw at Kissalala. There were only a few macaws
and parrots; but I was told that they had been
numerous just before I came, and were still to be
seen in large numbers on the plain towards the
lake. Besides these, there were also small birds,
mannikins, &c. Temple once brought me a very
strange-looking bird; its wings were very short in
proportion to its body; it had a lengthy tail, and
its general colour was a metallic copper. I am
sorry to say that I left this specimen, with others,
in the country, and have not-as yet received them.
Here, as is usually the case in the forests of Central
America, insects are to be seen in almost endless
variety.

I was in some measure confirmed in my belief
that the Woolwa tribes extended farther to the
west than Kaka; for one of the officers, who was
exploring the neighbourhood of the mines, said that
he had fallen in with a track in the woods leading
westward to a village of Indians, who, from his

description, resembled the Soumoo Woolwa in every respect.

Ere leaving Chontales, I sent Temple to Captain Pim, to give him such information concerning the river as his long experience would enable him to impart.

CHAPTER VI.

On the 23rd of March, in the afternoon, I left Santo Domingo for Kaka, on my way back to the coast. I managed, with much difficulty, to walk to the settlement, reaching it only just before dark, and very much fatigued. I mistook my road among the numerous hunting-paths made by the Indians. In such cases it is necessary to keep presence of mind. The sun was getting low; and having sent Temple on to make arrangements for the night, I walked leisurely along, till I discovered that I was on the wrong track. Remembering what very uncomfortable predicaments travellers have found themselves in under similar circumstances, by becoming first flurried and then hopelessly perplexed, I simply turned round, and retraced my steps to a point where I was confident I had been on the right path. After this, I continued my way more carefully, until I met two of the Indians, who were seeking me, alarmed at my long absence. Having refreshed myself by a bathe in the cool waters of the creek, and partaken of a good supper, with which my Santo Domingo friends had provided me, I spread my blankets, and lay down by the

SOUMOO OR WOOLWA INDIAN. BLEWFIELDS RIVER.
CENTRAL AMERICA.

fire once more, as I had been wont to do at the different Soumoo villages. I had suffered much from thieves during my short residence in the Spanish country; all the more, probably, as at first, from a habit acquired among the honest Woolwa, I took no precautions. The consequence was that now, when I left Santo Domingo, I had not even a strap with which to secure my blankets, and was obliged to use string instead. Poor old Temple lost all his clay pipes, and even his top-boots. I was much vexed to find amongst the missing articles a meerschaum, nicely carved with the figure of an Indian chief, which had afforded much amusement to the natives. Large robberies are not common with the Nicaraguans; but they are much given to this petty pilfering, which is very annoying.

On the 25th we began to drop down the river, now very low : I was only able to hire men from place to place. At this season of the year all the Indians are engaged in working on their plantations; the weather being dry, they burn patches on the banks of the rivers, and clear new provision grounds. The three following days, while awaiting the arrival of a pit-pan and men to pursue our journey, we remained at the house of an old lady with six sons.

Having spoken so much of the Woolwa Indians, it may not be amiss to add a description of their personal appearance. The men are generally rather under the middle size, muscular, but often rather squat, probably from being so frequently in their

pit-pans; the expression of their faces is usually
good-natured; the eyes are black, large, and brilliant, the nose prominent, and as a rule aquiline;
the mouth rather large, and the lips thin; the skin
is of a warm chocolate colour. The young men are
very handsome. As I mentioned before, they have
a custom of flattening the heads of their infants.
I frequently have seen babies with their foreheads
so compressed by the broad flap attached to the
head of the cradle, that their eyes seemed ready
to start out of the sockets. They do not, however,
seem to suffer pain, and I fancied them to be quieter
than the average of young children. The flattened
forehead in the adults would probably escape the
notice of a casual observer, on account of the thick,
heavy masses of hair that cover it. I am not aware
that the Rama or the Moskito follow this curious
custom, which, as practised by the Woolwa, is very
similar to that of many of the North American
tribes. The name, *Woolwa*, must, I think, have
originated in a nickname, as these Indians always
call themselves Soumoo, and the traders also make
use of the same word in reference to them.

A party of Indians arrived one evening with a
net; they were going to proceed up the river, to
drag in succession the numerous pools scattered
among the huge boulders of rocks now rising above
the stream, but covered in the rainy season by
raging and foaming torrents.

The Indians seemed to consider their work
rather dangerous, especially on dark nights, from

the number of venomous snakes which infest the neighbouring banks. In exchange for a bone-handled knife and a few steel beads, I obtained from one of the young men of the party some of the handsome bead-collars worn by the natives in full dress.

Close to the landing-place stood a very fine vine-grown tree, which was a great resort for small birds, parroquets, finches, fly-catchers, and the beautiful and conspicuous, though common, black and blue creeper *(Cœreba cyanca)*, that Temple informed me was only a visitor during the dry season. Indian boys secure small birds alive by stunning them with arrows, fashioned at the points into a broad, button-like head; the Klings of India, I believe, are in the habit of using pellets of clay from a "sum-pitan" for the same purpose. Several kinds of caterpillar were much dreaded by the natives; but these do not appear to be so virulent as the South African species spoken of by Chapman, which, if they come in contact with the skin, are said to cause irritation for months afterwards.

On the 29th I started again, with Temple, two men from the neighbouring village, and one of our hostess's six sons, whom the poor old lady was much distressed to part with; when he stepped into the canoe, she sat down, covering her face with her hands, and crying in the subdued manner which is peculiar to Indian women. The other two young men, who, it appeared, had been in the employ of Temple at Blewfields, said that it would take five

days to go from this river to the Cooringwass, partly
by way of a creek and partly through the forest.
This intelligence decided me—my ankles being still
unhealed—to give up the idea I had entertained of
entering Pearl Cay Lagoon by the Cooringwass, the
usual Indian route, as it flows into the northern
portion of the lagoon.

On our way down the river we saw a great
many of the pretty little pigeons *Peristela cinerea*,
flying generally in pairs; the hen bird is brown,
but I was unable to obtain a specimen.

At Woukee, where we passed the night, we
found that, in pursuance of their custom, all the
people had left the lodges on the high bank, and
were camped on the little gravel island in the midst
of the river below. Sleeping in the open air at
this dry season would have been delightful, had it
not been for the sand-flies and other creatures which
tormented us. We started in good time the next
morning, and passed the Second Hill before mentioned; but ere doing so we were obliged to make
several portages, owing to the sinking of the river.
Once we made no less than six, including the long
one over the hill; the black rocks were so hot that
I could hardly walk on them with my bare feet, and
standing still was quite out of the question. On
one occasion, I had noticed Temple, who was in
front, had gone a long way round, but thinking
that he must have been looking for something, I
took a short cut across what appeared to be a smooth
shore of white sand; I had scarcely reached the

middle, when my feet sank in, and I discovered that it was so intensely hot as almost to blister the skin, and by a series of desperate strides I arrived at the water's edge, and learned by this experience to be more careful in future.

That night we camped on a shingly beach, near a gigantic fig-tree. This is a very handsome ornament of the forest, but does not in the least resemble the cultivated fig; it has long trailing branches and gloomy leaves, and the large howling monkeys are very fond of it, generally frequenting its vicinity.

After dark I heard a shrill whistle close to our camp, which the men said was the tapir, or mountain-cow, calling to its young.

The next day we passed the settlement where I had seen a very pretty little parroquet as we ascended the river. I landed with Temple, in hopes of being able to get it for my sister, but only found women there, all the men having gone down to Blewfields.

When Indians arrive at a place where none but women are at home, they do not land, but occasionally hold communication with them from the canoe.

Although this was Sunday, we were obliged to travel some distance, as our provisions were fast diminishing. The birds turned out in endless flocks, as if they knew it was Sunday; so I shot a large bittern, off which we made a good dinner. We also came across a number of the beautiful

R

Muscovy ducks, but unfortunately my gun was not loaded. In the evening we saw a company of red monkeys among the bamboo, and being in want of meat, we put Temple on shore, that he might take one for supper.

On the 1st of April we started in good time, passing the First Hill (the longest portage), and camped on the rocks some distance below. On the upper part of these rivers there are many flat rocks, which I always chose as the most desirable camping places in this damp climate.

The river was now very low, and the Indians used after dark to take a brand from the fire in one hand, a fish-arrow in the other, and walk about the shallow margin of the stream, in search of cray-fish and crabs, which they secured by a lucky arrow-thrust, and then dropped on the embers to cook; they also ate the freshwater snails, cooking them in the same manner.

We passed the Moroding Falls about the middle of the next day, and again camped on some rocks, making our suppers off Woukee bittern.

The Woolwa are very skilful in the management of their canoes, especially in the more difficult parts of the river. In descending a rapid, one man stands in the fore, with a pole balanced in the middle; with this he touches the pieces of rock, right and left, as the canoe shoots along, thus warding off the prow when apparently on the point of dashing against them; another man (or the wife) sits with a paddle over the stern, and

assists with a timely stroke of the broad blade.
When we came to a rapid, we lightened the canoe
by landing Temple and one of the Woolwa. I
remained in the "pit-pan" with the other two, pre-
ferring this to a walk over the sharp rocks, so often
scorching from the heat of the sun's rays.

Soon after daybreak the next morning, we
passed the spot which had been our second camping
place in ascending the river from Kissalala. I was
surprised, when I sat down on one of the fragments
of rock near the fire, to find that its surface had
been evidently carved in some flowery pattern in
bold relief; for, although much water-worn, the
pattern, somewhat similar to that seen on the arms
and legs of European tables and chairs, was distinctly visible. Temple told me of high rocks
on the Rusewass that are completely covered with
figures of men and women, and other devices. He
also asserted to have seen on the same river a pool
among the rocks which, when the river was low,
continually boiled up, throwing out clouds of steam,
and killing all the fish within its range.

There is a small, yellow-breasted bird of the
family of flycatchers *(Syranindæ)*, about the size
of a yellow bunting, which is very common on this
river; it is called, in Moskito, *Kisscadee*, and lives
on insects taken on the wing, in much the same
manner as our English flycatcher. This bird looks
as meek as a dove, but I saw one attack a large
Woukee bittern, whose leg I had broken with a
shot. On another occasion, I saw several of them

maintaining a very successful warfare with one of the large crab-catching hawks. They have a great advantage in always keeping above their enemies, and, as they fly, striking down at them.

Jaguars, or " tigres," as they are called here, do not seem to be uncommon in the forests, but I never had the good fortune to encounter one personally. The puma is also seen occasionally. I remarked at one of the villages a Woolwa boy whose forehead had been totally disfigured by a blow from the paw of one of these tawny brutes. He had been left alone in the encampment while the men went out to catch fish. On hearing his cries they hurried back, and were just in time to save him, by driving their arrows into the puma's body.

In the afternoon we passed Kissalala, which place presented a most desolate appearance. The thatch already had been blown partly off the houses, and the whole land was choked up with weeds. I had heard from Kennedy and his son, whom we met on their return from cutting plantains in their provision-grounds, that Teribio and the rest of the Kissalala people had not come back to the settlement since their desertion on account of the "sickness," but were now encamped lower down the river. We fell in with them that afternoon near the mouth of the Rusewass, and they all seemed glad to see me again. Teribio said he would come to Blewfields in three days, to receive payment for going up the river with me, and gave Temple a knotted cord to mark the time, according to their custom. He was also to bring the young monkey I had entrusted to his care; it was now very tame. These animals carry their young on the back, and should the mother be shot, of course both fall to the ground together, and the little monkey is secured before it can regain the trees; but it will rarely leave the body of its mother voluntarily. Monkeys, when required as pets, are caught in this way.

We took our evening meal at the camp, and pushed on by starlight, passing the Rusewass and the Rama mouth; and not liking to lie down on the banks in the dark, for fear of snakes, we slept in the pit-pan.

I heard that on the Rusewass, besides the figures of monkeys, &c., there is some "writing" cut into the face of a rock. May not this be some inscription in Latin, left by the early Jesuit missionaries?—who, it is well known, in their untiring zeal, penetrated to the remotest quarters of the globe.

We started again on the morning of the 4th of April, about two hours before daybreak, which always presented an imposing scene on the river. A steamy mist generally hung over the surface of the water, and not a sound was to be heard, save the strokes of the paddles, and the occasional hoarse croaking of the tiger-bittern as he rose with heavy wing from the bamboo or reeds where we had disturbed him. We breakfasted at the usual halting-place, called Starhouse, which was the only suitable spot for making a fire for some distance. I here tasted the eggs of the iguana, of which the Indians had found great numbers, as this was the season for laying; in flavour, they are not unlike ducks' eggs: two or three dozen are generally found together. The iguana, alligator, and freshwater turtle all lay their eggs at this period, and bury them (much in the same manner) in the dry sand on the river's banks; but I was never able to conquer my aversion sufficiently to taste those of the alligator. The eggs are very artfully concealed, but the natives are equally clever at discovering them. When, from the appearance of the sand, they imagine it has been disturbed, they cut a long, slight wand, and thrust it down a considerable depth; should the point, when withdrawn, have some moist particles adhering to it, they examine and smell them, and, having come to a satisfactory conclusion, immediately turn up the sand until they find the eggs; as the shell is exceedingly hard, they are thrown into a heap, and then taken down to the canoe. In the latter portion of our journey, we often stopped for this purpose, wherever the quick eyes of the Indians discerned a favourable spot.

In passing a thick cluster of broken and entangled bamboo, we found a large *hicatee*, or freshwater turtle, wedged fast between the stems close to the margin of the stream. It had, no doubt, been disturbed by our approach, while on the bank for the purpose of laying its eggs, and had got into its present predicament in endeavouring to regain the water; it was wedged so firmly, that the Indians had to use their machetes freely before they could extract it, and place it in the canoe, where it remained helplessly on its back for the rest of the journey.

We met, as we advanced, two Woolwa pit-pans on their way home to Woukee; most of the women had bitterns, or puppies, which they had obtained at Blewfields.

Presently we came to a plantation belonging to Temple, pleasantly situated on a point formed by the junction of Mahogany Creek with the main river. Temple found his family there; so we went no farther that day, but had a very good supper with the people of the place, consisting of dried turtle and the usual plantain and cassada, to which I added a couple of bittern, shot during the morning. The meal was laid out on freshly cut banana leaves, and spread on the hard ground in front of the bamboo hut. On this day, for the first time after leaving Kaka, we had a shower of rain.

CHAPTER VII.

On the 5th I continued my journey to Blew-fields, in one of Temple's dorys, paddled by Melville, Temple, and another creole lad. Temple was to follow, as the pit-pan was too deeply laden to cross the open lagoon with safety. We stopped for a few minutes at another creole plantation belonging to Crawford, the man in charge of Dr. Green's house at Blewfields. They were engaged in boiling sugar-cane juice, and presented me with a bowl of it warm, which I found very good.

In the afternoon we gained the mouth of the river, where the country, once covered by a fine forest, now presented a still stranger aspect than when I saw it before in the month of November. Most of the shattered and decayed trees which were then standing had now fallen, and their bleached trunks lay piled up in shoals, like the whitened skeletons of immense animals.

We had had to paddle against a strong wind down the river, and had run aground in sailing across the lagoon by moonlight, and were delayed some time in setting the canoe afloat again; therefore it was late when we got to Blewfields. During

our sail we had passed many large fish, coursing
rapidly through the dark water all round, and
leaving in their wake phosphorescent light. I took
my tea of roast plantain, and enjoyed a pipe at
Temple's house, and then, wrapping myself in my
blanket, lay down on a bench.

Next day I went to the Mission-house, where,
as usual, I was most kindly received by Mr. and
Mrs. Lündberg, who put into my hand a large
bundle of letters, which had been accumulating for
a long time, and expressed their pleasure to see me
again in such good health.

I much enjoyed the simple Moravian service the
first Sunday after my arrival; it was still held in
the school-room, as the church was not yet rebuilt.

I put up my hammock again at Dr. Green's
house, and during the time I remained at Blewfields I often persuaded Crawford to go out shooting
with me early in the morning, before the sun rose
high. He used to bring round his dory to the little
jetty of stones before the house about daybreak.
We then sailed across the lagoon to some unfrequented part of the shore, and paddled amongst the
creeks and mangroves at the north end, or along
the inside of Deer Island, generally returning at
mid-day.

Besides the birds I shot myself, the young men
often brought me some of their own capture to skin,
in exchange for a fishing-line, a few hooks, and
other things of equal use to them.

After the lessons at the Mission-house were over, the dusky scholars sailed their boats along the shore, reminding me of the London boys on the Serpentine. The Blewfields urchins, however, are in the habit of disrobing to wade through the shallow water; and their boats, though merely home-made models of their fathers' dorys, generally sail much better than English boys' self-constructed vessels.

On the 11th, I went out in Crawford's dory for a sail among the little islands or cays in the lagoon, and along the shore of Deer Island, on which we saw many racoons. I also saw a curious bird, the boat-bill, called by the Indians "Cooper." The little green heron is everywhere common by the water's side, and is easily recognised, while yet unseen, by its loud cry of "tuk-tuk-tuk," from which it derives its native name. The noise it makes is loud enough to be produced by a much larger bird.

The coast canoes, or dorys, sail very well; in fact, Crawford said that he had once run sixty miles in eight hours in one of these boats. Mrs. Crawford, who was a fine specimen of the coloured people of this coast, showed me many little attentions. When I came in, as was often the case, during mid-day's heat, she would have a nice clean cloth spread in their cool bamboo house, with a bowl of soup, some bread-fruit, or plantain, and plenty of refreshing syrup, of which they here use a great deal. I would then go home to preserve the birds shot during the morning, and spend the evening with my kind friends at the Mission-house.

The dry weather still held out, and it would have been oppressively hot but for the strong north-east sea-breeze. The most tiring part of the day was before ten o'clock, until which time the sea-breeze did not set in.

On Sunday, the 14th, there was a very interesting service at the Mission-house—Baptism and the Confirmation of Adults.

I had once a conversation with Crawford on the subject of some Indian antiquities which were found in the neighbourhood; they consisted of broken pottery and stone hatchets (the former moulded in the shapes of heads of men and animals). These curiosities were found in large mounds of cockle-shells, when cutting away the jungle to form new provision-grounds, at a place called Cookra Point, the next headland in the lagoon to the south of which Blewfields is situated. A specimen in Mr. Lündberg's possession was a very good representation of the head of a wild hog, or warry. I went one day with Mr. Pinnock, the schoolmaster of the Mission, a native of Jamaica, for the express purpose of searching these cockle-mounds at Cookra Point, hoping to find some relics, but there was little to be seen but fragments of pottery, somewhat differing in shape and pattern from that made by the Woolwa Indians. These mounds are now completely overgrown with dense jungle, except in places where they have been laid bare in the making of provision-grounds for cassada and plantain.

Teribio, according to promise, came to Blew-fields, bringing with him the little monkey that we had secured at Kaka, and which he had educated for me; it was now the tamest and gentlest monkey I had ever seen.

I found it very difficult in this country to preserve the insects which I captured, especially the butterflies; they were all in constant danger from the ants, and in the wet season the mould and mildew were nearly as destructive. I was only able to procure a few land shells; they do not seem to be plentiful here, the creoles being unacquainted with the word " snail."

No little excitement was caused in the place at this time by the arrival of a new missionary and his wife, in the " Messenger of Peace." They were received by a large portion of the congregation, who assembled on the stone jetty singing hymns. It was a very pretty sight as the canoe approached from the Mission schooner, which lay some distance off, riding lightly on the moon-lit waters of the lagoon, that danced and rippled on the sand and stones, under the influence of the gentle sea-breeze. Two North Americans also came from Grey Town in a Carib boat to see the place, intending to extend their trip to the Corn Islands off the coast, which form part of the Moskito territory.

The Lenten services at the Mission were very

numerous, and there was always a remarkably large and attentive congregation.

A number of jaguars, forced probably from the woods by the scarcity of game since the destructive hurricane, had been committing great havoc among the goats, pigs, &c., belonging to the inhabitants. In order to put a stop to these depredations, traps were set in the most likely places, but for a long time without success. However, one night, an old lady, one of the last of the original white settlers, hearing a commotion among her live stock, ran out to see what was the matter; her surprise, no doubt, was great when she found herself face to face with a large jaguar. She did not lose her presence of mind, but flourishing an umbrella, the only thing she had in her hand, she suddenly opened it full in the animal's eyes; upon which he was so startled, that possibly desiring to escape, and seeing only one opening, he immediately sprang through the door of the trap, which closed upon him securely, and the next morning he was executed without difficulty.

Temple had been away for some time, but as he had now returned we made several expeditions together. One morning we tried to obtain some of the white cranes, but they were very shy, and remained too far off on the dry shoals about the river's mouth. We afterwards went to Cookra Point, where, when I was with Mr. Pinnock, I had heard in all directions the peculiar mellow cry of the cock yellow-tail *(Ostinops Montezuma)*. The plumage

of these handsome birds is of a deep russet-brown,
changing to black on the head and neck, and the
tail-feathers are of a bright yellow; the top of the
beak is coral-red, and the cheeks pale blue. They
are sociable in their habits, living and breeding in
flocks, and the branches of some favourite tree may
often be seen covered with their long pendent nests.
The difference in size between the cock and the hen
is very great, although the plumage is the same.
They are probably attracted to these plantations by
the quantity of ripe " panpa " and banana. I was
told that large flocks of a smaller kind were seen at
certain seasons of the year.

We shot six, but only secured two, in spite of
Temple's use of the machete, the undergrowth of
bush and wild cane being so matted together by
creepers and bush-ropes. In cutting through this
jungle, we came repeatedly upon those large mounds
of cockle-shells already mentioned.

After a fatiguing but an agreeable morning, we
repaired in the mid-day heat to the cool and shady
beach, where the canoe had been drawn up, and
regaled ourselves with green cocoa-nuts from one of
Temple's trees that grew near. I have often drunk
off three of the green nuts in succession when
thirsty, for a large quantity of this delicious and
slightly acid fluid may be taken without leaving
any of the unpleasant feeling which would supervene
after drinking a like quantity of water; and it is, I
think, at the same time, the most cooling and refreshing beverage that it is possible to take on a hot day.

We frequently went again to Cookra Point. A great variety of butterflies were visible in the thickets of sage and guava bushes in the rear of the settlement.

The half-yearly Congress now took place at the Mission-house, to be present at which the missionaries congregated from the different stations along the coast. Captain Pim arrived the same day from Grey Town; so that, with the two Americans and myself, there was quite a large party to dine at the hospitable table of Mr. Lündberg.

Captain Pim started that evening for the purpose of ascending the Woolwa river as far as Kissalala; he took an extra complement of paddles for his canoe, saying that he intended to make the quickest trip that had ever been made: and he certainly was speedy, for he returned to Blewfields at 7.30 in the evening of the 30th. His object was, I believe, to inspect the lower part of the river, in connexion with the proposed route to the Chontales mines.

Cheese from Nicaragua frequently finds its way down the Blewfields river to the coast, where it is much esteemed; it is brought down by the Woolwa, having been passed by them, from one hand to another, from the border settlements.

Considerable excitement was caused at Blewfields by news received that the Nicaraguans contemplated possession, of what remained of the Moskito territory. It was even mooted that an embassy should be despatched to Lord Stanley, in order to put in an appearance in opposition to that sent by the Nicaraguans. A public meeting was held on the 1st of May at the King's House, over which Captain Pim presided; the men attended in large numbers, and discussed the encroachments of the Nicaraguans, and the best means of stopping their further progress. Captain Pim promised that, as there was no person in England to plead their cause, he would defend them to the utmost of his ability. Some resolutions were also passed, not very favourable to the manner in which their officers had discharged their trust; and these were to be brought before the executive council, which was to meet in September, when the Moskito chiefs from the northern part of the country were to be present. The meeting broke up in the usual way, and the next morning Captain Pim returned to Grey Town.

I called one day, whilst out shooting, at the little cay on which Mr. Rhan, an Englishman, lived with all his family; but he had been unwell for some time, and was now quite laid up, so that I missed seeing him. The view from the highest point of this little island, which is called Cassada Cay, is the most beautiful in the neighbourhood: the ocean and Blewfields Bluff on one side, and the lagoon, with its diminutive cays, on the other; in the distance, the high blue hills behind Monkey Point form a prospect which is scarcely to be surpassed in loveliness.

Mr. Lündberg kindly offered me a passage in a canoe about to leave for the Mission Station at Pearl Cay Lagoon; which I accepted, thinking that it would be a good opportunity of visiting that locality before the rain set in. The weather during the greater part of May was beautifully fine, but very warm. On the 2nd of the month, having put together my guns, some ammunition, and other necessaries, I went to the little jetty in front of the Mission-house, where I found the large canoe waiting. We started about seven o'clock in the evening, and took nearly the whole night to beat round the bluff, as the tide was running in. After arriving at Pearl Cay Lagoon, and passing between the low, sandy shores at the mouth of this large sheet of water, we sailed across to the settlement of English Bank. The only prominent object on entering the lagoon is Cookra Hill, which rises beyond the savannahs that skirt the shore in that direction. We landed in the afternoon at the boat-house of the Magdala Mission Station, where the minister, Mr. Grunewald, and his good lady received me very kindly, and prepared a good cup of coffee, of which I stood much in need; for, in consequence of the boat having been full, I had not lain down all night.

Magdala is built much the same as Blewfields, but upon lower and more sandy ground, with a savannah behind and hills in the distance, but of no great elevation, Cookra Hill being the highest. It is so called from a tribe of Cookra Indians, who are said to live in the forest beyond; one of these, a settler in this place, was pointed out to me: he seemed taller and slighter than most of the Woolwa

s

Indians, and of rather lighter complexion. I heard
that a few members of this tribe had become "tame,"
and had settled on the northern side of the lagoon.
Many strange stories are told about these people,
who are represented to be as wild as deer, wandering
about the densest parts of the forest, which are
inhabited by the Woolwa, and never making a
lodge or canoes.

After dinner, Mr. Grunewald walked through
the village with me. At one of the houses, a man
had just returned from a successful hunt, having
killed two large manatee, one of which weighed
over a hundredweight. These animals bear some
resemblance to the seal, and feed on the long rank
grass growing on the banks of swampy rivers, creeks,
and lagoons; they are found in large numbers on
the northern side of the lagoon.

The next morning I took my gun and went for
a stroll on the savannah, and passed, at a short
distance, through a village of Moskito people of
mixed race. This place is called Hawl-over, being
situated on the lagoon, near the spot where the
canoes are hauled over to a creek when going by
the inner route to Blewfields Lagoon. A number of
cachew-trees grew wild all round: as it was very
hot, I quenched my thirst with the juicy fruit.
The walk in the open country was very enjoyable—
a change from the canoe mode of locomotion that I
had been obliged to resort to, on account of the
dense woods interrupting land travelling, with much
disadvantage. The fires occurring at the periodical

burning of the grass cause much destruction among the pine-woods upon the savannahs. The pine and the palm here grow together. One day, when shooting among the pines, I slightly winged two very handsome hawks *(Astur magnirostris)*, which I managed to secure, and carry home in the creel I had on my back. After having kept and fed them for more than a week, one of them made its escape, when its wound had healed; the other followed it the next day, at which I was much vexed, as they were beautiful birds, and in full plumage.

A very large jaguar had been committing cattle depredations at the settlement; and once, when out on the savannah, I saw the tracks on the sand, and judged from their dimensions that the animal must have been of unusual size. Mr. Grunewald had, I believe, lost twelve of his finest cows.

Birds are by no means numerous in this neighbourhood; but the cry of the *Psitorhinusmoris*, " Pean-pean," as the natives call it, from its note, is always to be heard on the savannah. It has a curious knob of skin at the base of the neck, which, I suppose, can be inflated at pleasure. In colour, it is a sort of dull drab, shaded underneath with white, and its tail feathers are tipped with white. In habits, this bird resembles the magpie, hopping on the ground and amongst the branches of the trees in the same springy manner. The " hen-hawk," as the creoles call it *(Astur magnirostris)*, is very common among the pine walks in the savannahs, and large green parrots fly in chattering pairs overhead morning and evening. Their flight is exceedingly rapid and powerful. Trogons, pigeons, and other birds were to be seen in the thickets round Hawl-over, and a red-headed woodpecker. I also shot a little bird, the *Attila citropygius*. The *Dryocopus scapularis* is common in the clumps of trees that are scattered at intervals over the savannahs, but, as I before observed, there are very few land birds.

A sluggish little stream wound through the savannah between marshy banks from the direction of Cookra Hill into the lagoon, just below the Moskito village of Ritepoora; it was full of fish of different kinds. Some I saw caught by a Moskito man, were rather assimilating to large perch in form, but spotted like trout. Hicatee, or freshwater turtle, and swarms of juvenile alligators, were to be seen with their snouts above the turbid water. The little green heron, or *tuk-tuk*, had its abode among the water-lilies, and the white crane was visible in the distance, but, being very shy, it would not admit of a near approach. Some large snipe, and sometimes, though rarely, a Muscovy duck, would rise from the sedge.

Mr. Grunewald very kindly put one of his canoes at my service, and engaged to accompany me a young man named Delancy, of very mixed race; indeed, these self-styled creoles are generally so mixed that it is impossible to discover which tribe they belong to, and therefore one must use their own name in default of a better.

to face page 260.

Tb. Wickham

Ritepoora, Moskito Village, Pearl cay Lagoon, 1867.

The traveller is here reminded of Christmas, as at home, by little observances kept up by the coloured creoles, and probably learnt by them from the stray English of the West India Islands from time to time settling at various points on this coast. Among others, may be mentioned that of dressing out a small tree in the pleasant month of May, after the manner of the old English Maypole of the village-green. I was also surprised at Christmas to hear Temple and the creoles talk of " egg-flip."

CHAPTER VIII.

MAY 10TH.—I accompanied Mr. Grunewald to the neighbouring Moskito settlement of Hawl-over, where he held a prayer-meeting once a week, and preached to the people in their own tongue. The scene, although so simple, would have made a good subject for a painter. The missionary standing, book in hand, in the centre of the roughly-thatched hut, surrounded by a circle of dusky listeners, the men on one side and the women on the other, speaking to them pleadingly in the sonorous Moskito language; whilst outside, in the sunlight, some thoughtless girls were laughingly peeping through the crevices between the palmetto, or *poptard*, stems, or leaning listlessly in their scanty but picturesque attire against the posts of the doorway.

Although, no doubt, in the old times the Moskito men were very superior in war to the Woolwa, Rama, and other tribes of the country, yet they do not appear to me at present to bear a very favourable comparison with them, mixed, as they have become in most of their villages (with the exception, I believe, of a few to the north, towards Sandy Cay), with former African slaves from Cuba and other parts of the West Indies.

There was an old African from Cuba at Magdala at the period of my visit, who, though he-had joined the Moravian Church, still wore a little wooden cross suspended from his neck, of which he seemed to entertain a great opinion.

I now went out shooting every evening around the lagoon, accompanied by Delancy. This youth had heard from some quarter that I was in the habit of sketching, or "taking pictures," and, immediately after my arrival, he and some of his friends wanted their "likenesses taken." I had been besieged in the same manner at Blewfields, having once thoughtlessly given a handsome young creole a sketch I had taken of him, which I suppose had been handed about; for afterwards nearly all the young men and girls sought every opportunity of making a request for "their pictures," often stopping me in the path, coming to me in my quarters, and hanging about until I asked them what they wanted, well knowing what the answer would be. This incident should be a caution to travellers.

About the shores of the lagoon were great numbers of a handsome rail, which the creoles call "*topknot chick*" (*Aramides Cayennensis*); it is very delicate eating, the flesh being milk-white. It has a habit of skulking under the reeds and bushes on the shores during the day, and sometimes, when congregated in marshy places, makes a great noise by chattering in chorus. When shot, it goes through more contortions than any other bird I know of, not running away when wounded, but invariably tumbling on the ground, kicking and fluttering about in the most violent manner. A large red-breasted kingfisher *(Ceryle torquata)* is very common in all the lagoons and the lower part of the rivers; and blue and white garlings *(Ardea cœrulea)* are seen on nearly all the shoals and creeks: these latter have an exceedingly graceful appearance when seen feeding in the shallows amongst the interlaced mangrove roots.

The oyster-catchers were very regular in their visits to the oyster-banks near the mouth of the lagoon, arriving when the tide was low (although the difference here was not very great), and going out again along the coast when the tide rose. In sailing over the lagoon, I occasionally saw water-snakes *(Hydridæ?)* swimming on the surface; probably bent on crossing the expanse of water.

The attendance here, as at Blewfields, at the mission services was very large in proportion to the size of the place; and I noticed that some Moskito people came from Hawl-over and Ritepoora to join the Magdala congregation. A Woolwa lad, named Ramong, used to sit on a seat near the door; he had a very intelligent face, although his figure was short and squat, and his expression was one of the most beautiful I have ever seen. The singing, which forms a great part of the Moravian service on these occasions, is very hearty, and, although not of the finest, might in some respects be advantageously imitated by more civilized worshippers.

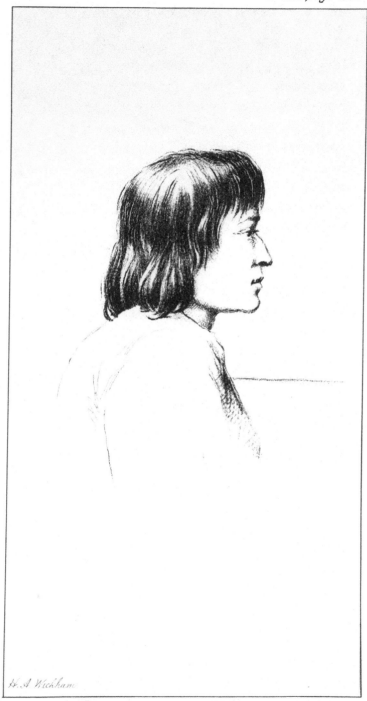

to face page 264.

RAMONC A WOOLWA.

H. A. Wickham

As the hymns were generally sung to the simple grand old tunes of Luther, the whole congregation was, without one exception, able to join, and the effect was better than that to be obtained by any of those paid choirs who select the tunes in order to show off their own voices, while the people stand and listen. Nearly all the teachers on this coast are from Jamaica. Some of the young Moskito women also came from the neighbouring villages to attend.

When they approached the precincts of the mission, they seemed to think it decorous and proper to remove the fastenings of their petticoats, usually worn at the waist, to the neck, so as to cover the upper part of their finely-rounded persons.

On the 14th, I went along the sea-beach, by a little lagoon and creek, among the mangroves, in quest of a flock of scarlet birds called in Moskito *powra*, which I imagined, from the description given me, to be the scarlet ibis, said to frequent this region. However, I saw nothing of them, and only brought back, as the fruit of my expedition, three bittern and a few other water-birds. Some large snipe were to be seen here, and active little sand-pipers ran in great numbers along the hot sands which enclosed the lagoon. I shot on the savannah a curious little goat-sucker *(Chordeiles texensis)*. One day, Mr. Grunewald accompanied me to a place called Rocky Point, on the shore of the lagoon, where most of the creoles at English Bank, and the Moskitos at Ritepoora, have provision plantations. We saw a great tree covered with the large hanging nests of the yellow-tail; but the young were fledged, and the flock gone elsewhere. A little to the north of Rocky Point is the Moskito village of Cockabilla.

I made another expedition after the "powra" from the same place, starting before sunrise; but this proved, like the first, a failure.

The creoles and the missionaries on the coast grow a large quantity of rice for their own consumption. Cotton, chocolate, sugar-cane, and pineapples are indigenous; and the Indians, as well as the coast-people, raise quantities of plantain, banana, and cassada: indeed, the provision-grounds return more of these productions than their owners require. The principal want is a certain supply of flesh. The beautiful bread-fruit tree, introduced from the South Seas, is now to be found at the Indian settlements far up the river. One tree I noticed at Woukee, but they did not seem to make much use of it. The cocoa-nut was in former times very abundant, though in the southern parts of the territory they had been destroyed in great numbers by the hurricane of 1865; and when I was on the coast, the people had just commenced to replant them. Once, many of the inhabitants depended entirely upon them for subsistence, handing over the fruits to the small American trading vessels in exchange for necessaries. The kind of life these people led may be imagined from the following saying, that "it mattered not what they did, for as they lay smoking in their hammocks, night or day, they could hear the sound of the falling cocoa-nuts, which told them money was accumulating, do what they might." What must have been their feelings, when, on the morning after the hurricane, they found the houses swept away, and their beautiful trees, on whose produce they reckoned for support, all broken, or their tough stems standing like bare poles, the feathery fronds twisted off and wrenched away by the rushing winds.

A little oil is made from the cocoa-nut, and also from the nut of the eboe tree. These people never eat the ripe cocoa-nuts raw, but use them principally for fattening their pigs.

About this time we had some very heavy showers, and on the 18th, while on the lagoon, I was caught in the heaviest rain-storm, with thunder and lightning, that I ever remember. I shot a species of night-heron (*Mychcorax violacea*), which the natives call "the carpenter": it is one of the few birds which they take the trouble to shoot for eating. It was fine again the next day, and I was able to have a little sport at the south end of the lagoon. All round the shores were large flocks of active fly-catchers, called the "wees-bird" (*Tyrannus intrepidus*), and a small falcon (*Tinnunculus sparverins*) is often seen perched on the tall pine-trees, or winging its rapid flight across the savannah in chase of the birds on which it preys.

I now determined to make a trip to the north part of the lagoon, and started on the 21st in a dory, accompanied by two men and an elderly creole named John Fox and young Delancy. As we sailed along this sheet of water, I could better appreciate its size, especially as the men informed me that it would take a whole day and part of the next night to reach the north end. When we started, the weather was fine; but the sun had not ascended far into the heavens before black thunder-clouds rolled up from the north-eastern horizon, and soon covered the whole expanse of sky, while the lightning flashed vividly, and the distant thunder muttered, giving us warning, during the stillness that intervened, to prepare for the coming storm. Presently we saw it approach us in the shape of a white squall, a well-defined line of white mist, raised from the surface of the water by the violent wind and pelting rain. The men had just time to take in the sail before it burst upon us, when we could do nothing but shelter ourselves as best we might, until it had in some measure abated. Then we continued our course, and leaving Tasprapowuee, where the Moravians have a mission station, to the north-west, we entered Slapping Creek, about two or three o'clock in the afternoon, by which time we were glad to perceive a slight change in the weather. The evening being fine, we took in the sail and paddled for some distance up the creek, to shoot something for supper. I saw here, for the first time, the beautiful heron called by the Moskitos "*Marara*" (*Ardea agami*); it was sitting quietly, the glossy deep green leaves of a shrub forming a good background to its graceful form, and appeared very tame, as if the bird creation in this remote creek was not often disturbed by the presence of human beings. Old Fox whispered that he heard warry in the neighbouring thicket, so I left the heron in peace. The banks of the creek being too damp for an encampment, we returned to the mouth, near which we found a suitable spot among some *Secumfra* palms. The long leaves of this palm, planted into the ground and supported against poles, form the usual mode of shelter for the Indians of this country. A number of the broad *waha* leaves were then cut and placed under this awning, and on them we spread our blankets. We had hardly completed our preparations, and had not even had time to light a fire, when the rain came down so furiously that the usually sufficient shelter only partially protected us. This continued for a space, while we sat on our blankets under the palm leaves; but at last it abated enough to allow of our making the fire, which we were able to light by means of the dry under-bark of the palms. I then made some tea, but the men preferred for themselves a kind of grass, called by them "fever-grass," of which they had brought a supply: it was prepared by boiling a pot of water, into which, after taking it off the fire, they threw a handful of the grass, and let it stand for a few minutes. I tried some, and found it very palatable; the flavour is not unlike the smell of lavender. While we were at supper, the rain began to fall again in heavy sheets, so as soon to fill the canoe, which was hauled up at our feet; it made its way through the palm-leaves above us, wetting us thoroughly as we lay in our blankets, and continued all night and far into the morning, while the thunder rolled and the lightning blazed; consequently, it was nearly the middle of the day before we were able to leave our camp for a little shooting in the creek.

I shot several of the beautiful "*Marara*," which I skinned and dressed on the spot. The evening turning out fine, we again paddled up the creek, which literally swarmed with Woukee bitterns, boatbills, darters, and other water-birds. It was surprising how they could find food enough in so small a space. As we paddled along, the bushy trees appeared to be alive with the odd-looking boatbills, fluttering and flying out in all directions, seemingly convulsed with hysterical laughter. There are two kinds of curassow: the more common is black, with a white belly; the other, called the *Queen Curassow*, is checked all over, in much the same manner as the tiger-bittern. It is an exceedingly handsome bird when seen in the woods, and erects its elegant crest most gracefully as it utters its deep note. A pair of the pretty little russet-brown "*jacana*," with lemon-coloured wing-feathers, kept flying in front of us, as we proceeded up the creek, alighting from time to time on the floating grass which covered the water near the bank: owing to the immense length of their toes, they were able to support themselves on this. I saw also several of the mud-hens *(Aramus scolopaceus)*, esteemed as a delicacy by the creoles. Darters were breeding high up the creek; their downy young being generally seen in pairs in a nest formed of sticks, usually placed on a branch over-hanging the water. They dropped out as we approached, diving and swimming about very actively; but whether they were able to return to the nest afterwards, is more than I can say. The darter seems to have much difficulty in keeping its balance when perching on trees, the feet being placed on the body considerably behind the point of equilibrium: this formation gives them great power of swimming under water, but makes them look awkward when out of that element. The neck is long and snake-like, and the beak curiously serrated, and admirably adapted for seizing fish beneath the surface. The eggs are bluish-white, with rather a chalky shell, small for the size of the bird, and are considered good eating by the creoles.

The night proved fine, and we passed it very comfortably, sleeping in the dory, shoved off-a little from the land, out of the way of mosquitoes. The next morning we went up the creek early, and I shot several birds, some of which I skinned as we went along, and the rest I finished after our return. The men struck two of the large fish called " *snook*" with their harpoons. After proceeding some way up the creek, we were obliged to flatten ourselves along the bottom of the canoe, to pass under the trunks of large trees which had fallen across. Two of the trees on the bank were inhabited by the yellow-tail, which were constantly flying in and out of their singular nests. On our way down the creek we had heard warry; so loading our guns with buck-shot, we went on shore, going in single file, as the under-growth was thick, old Fox leading the procession. When we came upon a drove, he fired a shot, bringing one down; while Delancy and I saw nothing of them, only hearing their rush as they charged away through the thickets, and the groans of the wounded animal, which was soon despatched, and carried down to the canoe. We then proceeded on our way down the creek; but near the mouth we heard sounds indicating the vicinity of another drove; and after having, with some difficulty, succeeded in forcing an egress through the thick green wall of weeds on the edge of the stream, we managed to land, and found ourselves in a comparatively open forest of secumfra palm and tall trees. The soft, damp mould underneath them was cut up with numerous footprints of the warry, until it was almost like the soil of an English pigsty. Being determined to have a shot this time, I kept abreast with Fox, and we presently came in full view of the whole drove as they were feeding: they soon perceived our presence, and rushed past in a body, thus giving us the opportunity of a splendid shot; at which old Fox became so much excited, that he jumped on a fallen tree straight in front of me (at the very moment I was about to pull the trigger), and fired, bringing

down one of the wild-hogs. It was quite provi-
dential that he did not receive the contents of my
barrel in his back. When I told him of his escape,
he said he was very sorry for spoiling my sport,
but " he had forgotten I was there."

On our return, we left Slapping, and landed, early
in the afternoon, on the narrow sandy shore, about
halfway down the lagoon, in order to cook a *snook*
for dinner, which we ate as we sailed along. We
arrived at the Mission-house in the evening, and I
set to work the next morning to finish the birds
I had partly dressed under such difficulties during
my expedition.

The little animal called " *quash* " by the creoles,
and " *coati* " by the Spaniards, is sometimes seen in
captivity in the Indian lodges; it is somewhat like
the racoon. The nostrils are arranged at the end of
its long snout in such a manner as effectually to
prevent earth and sand from getting up the nose
while it is grubbing for worms, roots, &c.; this
snout is exceedingly muscular, pliant, and sensitive:
the creature has a curious way of protecting it
from a blow or threatened injury by putting down
its head, and covering the snout carefully with its
fore paws. The arms and legs are stout and strong,
and the feet are armed with claws like those of a
miniature bear. The habits of the tame " *quash* "
in my possession, which now runs about the house
like a cat, are very droll and interesting; it has
formed a strong attachment to the little spider
monkey, and they never seem tired of playing and

T

frolicking together, their principal point of disagreement being that Quash is generally sleepy during the daytime, and Jacko takes a mean advantage of this, and pulls him most unmercifully about by his long bushy tail, only to be disturbed in his turn as he nods and dozes in front of the fire after tea, by which time Quash has become very sprightly, and bustles about the room with an air of busy importance, carrying his bushy tail straight behind him, with a gracefully undulating movement. While at the Blewfields Mission-house, Quash was a source of great amusement and some trouble; he was very friendly with all the dogs, and, unless securely shut up, on Sunday he would invariably follow Mr. and Mrs. Lündberg to the service; and on one occasion, when unable to do so, he got into the balcony opposite the church, and, having perched himself on the extreme ledge, made such a disturbance with his peculiar cry that some one had to be despatched to take him back.

CHAPTER IX.

On the 30th, I accompanied Mr. Grunewald on his visit to Hawl-over, as I wished to take a sketch of the old *Soukier*, one of a class of men who hold much the same place as the "medicine men" in North America; but he pleaded sickness, and would on no persuasion come out of his little hut, in which he continued to sit in such a manner that I could not obtain a glimpse of his face, although I stood by in waiting for some time. On a previous day, however, I had seen him talking to the men as I passed through the settlement: he was tall and meagre, with white hair, and altogether very interesting looking.

The medicine man, or *Soukier*, is a person of considerable importance in the settlement to which he belongs; he is physician, snake-doctor, and, as far as I can understand, priest: he goes through a series of grotesque incantations over a sick person, and after this has gone on for some time, he pretends to grapple with, and ultimately to secure, the spirit of the sickness. These people have acquired a valuable knowledge of the herbs of the country, which they have probably turned to account in devising simple remedies, and they are particularly skilful in their treatment of snake-bites. Among the Moskitos, the sick people are often banished to little temporary shelters, where they are attended by the *Soukier*. On one occasion, at the head of the Blewfields River, we met a *Soukier* posting down the stream in haste, having been summoned to attend some sick at a settlement low down on the Cooringwass: this shows that those of the profession who have established a reputation are often sent for to a great distance. I may here mention, among the insect plagues of this country, the jigger, as it is called in the West Indies, and by the Spaniards "*nigua*"; it is very minute, and generally buries itself in the foot, principally near the toe-nail, where it makes a nest, laying a great number of little white eggs, enclosed in a bluish bag. If not then extracted, the eggs will hatch, and each young jigger sets up on his own account, when the results may become very serious. I have seen boys at the settlement at Pearl Cay Lagoon whose feet seemed covered with warts, and their toes quite out of shape, from the number of jiggers they contained; this, however, was chiefly owing to laziness, and a want of care and cleanliness in neglecting to wash the feet often enough. The nest should be taken out with a needle or a sharp penknife as soon as its presence is known by the itching; this being done, no harm will follow, even should the bag be broken and some of the eggs left in, if the simple precaution be taken of filling the hole with the ashes left from a long-smoked tobacco pipe. The first time I made the acquaintance of these pests, I did not notice the slight itching, considering it only a small thorn, that would work itself out. Having let it remain for about a fortnight, I was one night kept awake by most acute pain, and, on examining the toe in the morning, I found it very much inflamed; and it then struck me what might be the real cause of my uneasiness. I extracted the eggs, and being advised to try the tobacco-ash, I did so, and it succeeded perfectly.

Butterflies were very numerous about the flowers of a bush which hung over the lagoon shores, called by the natives May-blossom; but there had been such heavy rains of late that the blossoms had dropped off, and when I provided myself with a net to catch the butterflies, none were to be seen.

The shores of these lagoons are, for the generality, lined with a thick growth of red and white mangrove. The appearance of these trees is so totally different from any we are accustomed to see in Europe, as to strike the observer at first sight with wonder; the curved roots that support the trunk rise in interlaced arches directly out of the salt water; and among these the waves of the advancing tide ripple and dance continually, causing a variety of curious sounds. These roots are covered with clusters of the little mangrove oysters, and at low tide peculiar little crabs run over the soft, black mud. The boughs which the mangrove throws out, thick, with glossy green leaves, are a great resort for king-fishers, herons, and boat-bills.

In the beginning of June, I left Pearl Cay, on my return to Blewfields; the morning was beautiful, the day dawning just as we arrived at the mouth of the lagoon. My Woukee bittern, the same bird which I had taken from the nest, and which had now grown very handsome, gave me much amuse-ment on the voyage. No sooner did we pass the bar, and find ourselves in the long, tranquil swell of the open sea, than he began to show symptoms of sea-sickness, being unable to sit upright, and twisting his long neck about in the most grotesque manner, with the evident intention of bringing up his last night's fish supper: at last he "went below," under one of the thwarts.

At this season, immense numbers of turtle pass along this coast in a southerly direction, in order to deposit their eggs in a certain locality of Costa Rica; there are two kinds, the common turtle and the hawk-bill turtle: the shell of the latter furnishes the tortoise-shell of commerce. In former times, the creoles made large sums by selling these shells at Grey Town, but the demand having fallen off, they scarcely take the trouble to hunt them for this pur-pose. We found a common turtle stranded on the sands; one of its fore-flippers and shoulders had been bitten off by a shark. It is a proof of the enormous power which these ferocious fish have in their jaws, that they can bite through the hard enamel-like covering with which the turtle is encased.

The brown pelican (*Pelicanus fuscus*) is commonly seen in small flocks upon the coast and lagoons, engaged in fishing, or, with a steady powerful flight, pursuing its way to more favourable localities. Its mode of fishing is curious : the bird soars upon his broad wings to a considerable height, and then, as soon as a fish is descried, it descends, beak foremost, upon the water with a sudden wheeling evolution, and with a force that would seem to dislocate its slight neck; seldom, however, failing to secure its prey. At other times, the pelicans may be seen swimming like geese in the shallows, composedly spooning up the shoals of fry with their capacious beaks. The quantities of fish consumed by them must be enormous. Occasionally, a solitary individual may be visible, perched, apparently in contemplative mood, upon a convenient mangrove bough. Alligators of very large size sometimes bask on the sands that fringe the swampy slips of land dividing the lagoons from the sea.

There was a great profusion of cocoa-nuts on the shore between Pearl Cay Lagoon and Blewfields Bluff, round which point we sailed, after a tedious passage, and entered the lagoon by the mellow light of a beautiful sunset. Tradition says that the buccaneers held Blewfields Bluff as a stronghold, and established themselves there, where a well and traces of fortification are said to be still visible. Here, those sea-rovers of old times allied themselves with the Moskito Indians, bade defiance to the Spaniards in the days of their greatest power, and

stored the treasure which they took in their forays
among the rich galleons on the Spanish main. The
place, indeed, looks an appropriate locality for such
scenes.

The weather in the beginning of June was unusually fine and dry for this humid coast. I was
rather surprised to meet at Blewfields with one of
the American gentlemen mentioned as having gone
on a trip to the Corn Islands, before I went to Pearl
Cay Lagoon. Mr. Lane was a regular Yankee of
the best class : his costume in this tropical country
consisted of a very ample brown-holland coat,
nearly down to his heels, and an immense straw
wide-awake hat, under which appeared his shrewd,
good-tempered, and generally laughing eyes. We
soon became great friends. I rather liked his honest
oddities.

On the 8th of June we agreed to go together
on a visit to Rama Cay, a mission settlement of
Rama Indians, situated on a little cay in the lagoon,
about eight miles south of Blewfields. After a
pleasant sail down the lagoon in a fast dory belonging to old Temple, we landed on the cay, which is
nearly covered by the neat little dwellings of the
interesting inhabitants. I had been the more inclined to go as the Rev. Jeus P. sen Jucrgenson
had very kindly invited me to pay him a visit when
I met him at Blewfields. He and his wife received
us very hospitably, and accompanied us round the
cay, showing us what would interest us, and calling
on the principal inhabitants, of one of the most

important of whom I took a sketch. This missionary must be a most active and energetic individual, the whole of this branch of the tribe having become Christians since his arrival among them, and the majority apparently very sincere. Most of the buildings in connexion with the Mission had been erected by his own hands. The construction deserving of remark for the greatest share of labour was a wide and deep well, the sides of which were beautifully built. The aspect of the whole place and people was an example of what can be accomplished by one single-hearted and devoted man. All this little island being occupied by the settlement, the people have their plantations on the mainland, and up a considerable river, which falls into the lagoon opposite the village. The number of inhabitants at the time of my visit was 164 (all Ramas), of whom 37 were communicant members of the Moravian Church. Previous to their conversion, this tribe appears to have fallen into a more depraved state than any other in the country; but now all was changed. The Rama race must have been numerous formerly, but only a remnant remains south of the lagoon and on the Rama and Indian rivers. I was told by an old Moskito that a small tribe had migrated south of Grey Town, and was now settled in Costa Rica.

In the evening, Mr. Lane and I sailed back to Blewfields, but the wind being contrary, we had to paddle round each point, and then sail across the intervening bights; consequently, we did not arrive till late.

I was very much vexed to find one evening that the fine Woukee bittern about which I had taken so much trouble, had been killed and eaten by one of the half-wild hogs that ran about the place, and found their way into the grounds surrounding Dr. Green's house, in search of bread-fruit fallen from the trees. I had intended bringing it home for the Zoological Society's Gardens.

I was surprised to meet one day, near Temple's lodge, a handsome young Woolwa, who had been one of the crew of my pit-pan on the river. He had a heavy axe (for the use of which these Indians are famous), and was engaged in cutting some logs of wood, to be used, I believe, in building Temple's new house. I was shocked to see how altered he had become; his skin, once as clear as bronze, was covered with rough blotches, the perspiration was running down in streams, and he seemed much exhausted. When with me he could not speak a word of Moskito; and I fear he must have had a hard time since then, for the creoles are inclined to be tyrannical, and make perfect drudges of the Indians when they have the chance. Had I known that Temple would have brought him to Blewfields to make a servant of him, I should have seen that he returned to his home up among the rapids, before I left Temple's plantation at Mahogany Creek.

The next day I started for Grey Town, to meet the homeward mail. Mr. Lane and I, after wishing our kind friends at the Mission-house farewell, put off, on the night of the 12th of July, to join the "Messenger of Peace," which lay in the bright moonlight under Halfway Cay, waiting for the breeze. As we sat on deck, watching the gently-heaving and silvery surface, we exchanged some of our experiences of the tropics, and a German, who had been long settled at Blewfields, and was now on his way to Grey Town on business, related, among other things, the sensations of a drowning man. He said that once, when crossing the dangerous bar at Grey Town in a small boat, it was capsized in the surf, and as he was not a strong swimmer, after making great efforts to reach the drifting boat, to which his companions were clinging, he began gradually to sink. He felt then as if he were suspended half-way, and was neither able to rise to the surface nor to fall to the bottom. When at length he did feel his feet strike the sand, he made a desperate effort to gain the dry ground by walking on the bottom, his great anxiety being now to keep his footing, for he felt that if he once stumbled he would never be able to rise again. The motion of the water swayed him from side to side, but still he managed to take many steps towards the sandbank, and already felt himself approaching the bright surface when something gripped his side. Thinking that it was a shark, he became insensible, but on coming to himself, found he was again in the boat, which had been righted. One of the men, a good diver, had gone

down after him, and it was the grip of his pre-
server's hand he had mistaken for a shark.

Next day, we had a very pleasant sail along the
coast. Between Monkey Point and Grey Town
the country is very mountainous, and the effects
of the heavy rain-clouds, now gathering for the
wet season amid the hills and valleys, were very
fine.

The Blewfields man in charge of the "Messenger
of Peace" was a brother of Hercules Temple. All
the conversation now was about the threatening
news from Nicaragua, and the captain loudly de-
plored the falling off of the warlike spirit of the
Moskitos. They were once sole masters of the
coast as far south as the San Blas Indians, who were
alone able to withstand the onset of their dorys
of war. But he expressed a hope that, if the hated
Spaniards did come, they would again clean and
sharpen their rusty old lances, and arise from the
drunkenness caused by the villainous stuff sold
them as rum by the traders, which, with the aid
of their own mishla, caused such demoralization,
in spite of the efforts of the missionaries. As we
sailed along, he pointed out several places where
these Indians had fought with those who had
formerly attempted to dispute their authority, and
related how the king used to go in his large dory
to take tribute of the Spaniards of Grey Town.

We reached our destination, and landed on the
morning of the 14th; and in the evening, returning
from a stroll in the woods, I met the funeral of a

Spanish child. It was a strange sight:—first came a lad with a spade, after him followed two men bearing between them the coffin, which was gaily painted and dressed with flowers; on one side walked the men, and on the other the women, some of whom were smoking cigars; behind, were men playing on fiddles and guitars; and, lastly, a number of people, who were throwing crackers about, and amusing themselves in various ways. One of these combustibles fell amongst a flock of guinea-fowl, which seemed to cause much merriment for the company.

During our stay at the Union Hotel, our charge was two dollars a day. Besides Mr. Lane and myself, there were several people there; one of whom, a gentlemanly young American, from Iowa, gave me a kind invitation, should I ever visit the States.

The Royal West India Mail Ship, "Tamar," left Grey Town Bay at noon, and steamed away for Colon (Aspenwall). Leaving the high mountains of Costa Rica to the west, we entered Navy Bay about the middle of the next day. The mountains behind Porto Bello looked very beautiful: they were the deepest blue imaginable,—here and there intercepted by dense rain-clouds and showers. As soon as the "Tamar" was secured alongside one of the wharves, I went for a stroll through the town. Aspenwall is built on the marshy island of Manyanilla, with mangrove swamps all round; the Panama railroad runs right through the town, and the odd-looking engines are constantly running to and fro. Seeing

some Indians engaged in selling their canoe loads
of plantains, sea-shells, &c., I mixed with the crowd,
and made some inquiries about them: they proved
to be a few of the famous San Blas Indians, who
had come from their village in the neighbourhood
of Porto Bello: the greater number of them seemed
to speak English.

We lay at Aspenwall for six days, which gave
all the time necessary for the little sight-seeing
that was possible where so little is to be seen. The
American Episcopal Church is a handsome building
of stone; but there was no service held in it during
my brief sojourn in the town.

I was exceedingly sorry to hear that Captain
Hill, by whom I had been so kindly received in
Chontales, had died on board the "Tamar" while
she lay at this place on her previous voyage. He
had been buried at a spot called Monkey Hill, a
little distance along the Panama Railway line. I
walked along this line one day for some distance,
and discovered that the first part runs through a
dark mangrove swamp; as far as I went there was
very little elevated land. The amount of human life
sacrificed in laying this line must have been immense, in consequence of the workmen turning up
the slimy deposit of ages under a fierce sun: they
say at Aspenwall that one man died for every
sleeper.

It was now intensely hot in the middle of the
day,—hotter, I think, than I had ever felt it
before, for though my skin was well accustomed to

heat by this time, the back of my neck became scorched and blistered.

Mosquitoes were very troublesome on board when we lay alongside the pier, and there was very little rest to be had below at night. I wondered how the natives could exist in the houses I had seen in the vicinity of the town; they were often built among the mangroves, on the little spaces where the swamp has been artificially filled in, and were only to be approached on boards raised in piles above the ooze.

There were two North American gunboats here; one was called, in singular bad taste, the " Osceola," after the celebrated Seminole chief, who died in a Yankee prison.

We left Aspenwall on the 25th, and I am sure every one on board rejoiced to see the last of it.

We passed Porto Bello, and coasted the country of the San Blas and other independent tribes; and on the following day we sighted the high land of South America, in the neighbourhood of Santa Marta; and at St. Thomas' we changed ships, and sailed for Southampton.

REPORT

ON THE

INDUSTRIAL CLASSES

IN THE

PROVINCES OF PARÁ AND AMAZONAS,
BRAZIL,

BY

JAMES DE VISMES DRUMMOND HAY, C.B.,
H.B. MAJESTY'S CONSUL AT PARÁ.
SEPTEMBER, 1870.

U

REPORT *on the* INDUSTRIAL CLASSES *in
the Provinces of Pará and Amazonas, Brazil,
16th September,* 1870.

THE population of the two extreme Northern Provinces of Brazil, Pará, and Amazonas, is composed
of several distinct classes or races of men, namely,
the Tapuyo, or civilized Indian; the white man,
descendant of the Portuguese; Europeans, and
foreigners from all nations; the negro; and, lastly,
the several lineages which have sprung from the
free mixture of all these races, and 'amongst whom,
especially in the lower orders of the towns, black
blood appears to predominate.

The immense area included in the two provinces
of Pará and Amazonas may be roughly computed at
about 400,000 square miles; and the existing population at not more than 350,000.

The population of slaves is small as compared
with other provinces in the Empire, hardly
amounting to 20,000. Their number is yearly
decreasing, for nowhere in Brazil is the feeling that

the abolition of slavery is necessary and imminent, greater than in these provinces.

The proportion of available labourers is very small, probably about one-tenth of the population; as, though there is some poverty, there is little suffering; for nature is prodigal, though men are inert; yet, from this very cause springs the want of labourers, in the field and town severely felt. Labourers and workmen are clamoured for, enabling the few to make up, to a certain limit, their own daily and often exorbitant stipulations for wages.

The European who emigrates to this country, and by temperate habits becomes acclimatized, competes successfully with the native,—suffering no further ills from the climate than others of the white or dark race who are born and bred here.

The Negro, or Mulatto, whether as free man or slave, is especially serviceable; but the apathy of these races enables the European, with his superior energy, to take precedence. Of all races, however, which appear in these provinces, the Portuguese, as a more cognate people, are the most successful. In the two provinces there are about 20,000 Portuguese; and in the capital of Pará, out of a population of 35,000 inhabitants, there are not less than 5,000 Portuguese. These are engaged either in commerce, or occupy such trades as shopkeepers, smiths, carriers, drivers, and boatmen—to the utmost total exclusion in these trades of every other nationality; and this, owing not only to their numbers, but to their speaking the language of the country, having the same customs, adapting themselves more readily to the food of the lower orders; and also, it must be admitted, to their general hardworking, sober conduct, and the clanship kept up amongst them.

The labouring agricultural population is principally composed of the Tapuyo Indians, upon whom falls, in a great measure, the burden of domestic employ; as also in meeting the Government demands for recruits in the national service.

The artisan workers, are a mixture of all nations—in which, Germans and Portuguese appear as carpenters, joiners, and masons; Englishmen and North Americans as engineers and mechanics.

All these classes have employed and set an example to the native races, who learn with aptitude the several trades or professions.

In all clever workmanship, the Mulatto, whether half Indian, or half white man, has shown his capability in an eminent degree by his general intelligence, sobriety, and attention to his work; yet does the native workman require kind and encouraging treatment, and refuses to be driven, or to be spoken to harshly at his work; for he then avails himself of the scarcity of workmen to resign his engagement, and seek employment elsewhere.

As has been already made known in my "Annual Report on the Trade and Commerce of Pará and Amazona," the extraction of india-rubber, which is here so prodigally yielded by nature, and requires little skill and experience from the labourer, has absorbed all attention over agricultural produce. The sugar-cane is only allowed to grow for the manufacture of cachass, or white rum, and sugar is imported from the Southern Provinces. The cultivation of cotton, rice, coffee, and many other products, is totally neglected, though known to be yielded by the soil in abundance and excellence; and though every facility is afforded for transport by the numerous rivers and water-ways which intersect the land.

The labour of extracting rubber is so small, and yet so remunerative, that it is only natural to mankind, and especially to the yet sparse and comatose population of these provinces, to prefer that occupation; in which a family gang, or single man, erect a temporary hut in the forest, and, living frugally on the fruit and game which abound, and their provision of dried fish and familia, realize in a few weeks such sums of money, in an ever-ready market, with which they are able to relapse into the much-coveted idleness, and enjoy their easy gains until the dry season for tapping the rubber-tree (June to January) or collecting nuts returns; rather than, by a little exertion, to clear the forest, till the ground, and watch their crops for a few months, before they can obtain a return for their expended capital.

An expert and steady Tapuyo (the class chiefly employed in extracting rubber) will collect about

to face page 295.

FRUIT (3) IN POD

FRUIT

H. A. Wickham

LEAF AND FRUIT OF THE CIRINGA TREE (Indian Rubber) SIZE OF NATURE.

eight pounds English per day, which on an average is worth 8 mil ries, or about 13s. 4d., i.e., 1s. 8d. per pound.

In a good rubber district men are known to extract even an arroba of rubber, or thirty-two pounds English, per day; but I have given the quantity of eight pounds of rubber as the average collection per man. Sometimes the Tapuyo, as also others, will lend their services to the proprietor of a district forest, in which the rubber-yielding portion is calculated by walks leading from tree to tree, extract rubber for their employers, and receive a per-centage on the amount collected, according to the value of labour at the time being. The method of extracting the milk from the rubber-tree is primitive; and still more primitive and rude is the manner of smoking, or curing the rubber-milk over smoke arising from a funnel, under which is fired an oily nut, which is the fruit or seed of palm-trees, the *Maximiliana Regia, Attalea excelsa*, &c.

Already, however, at this very time, an American and an Englishman have simultaneously presented for approval simple inventions, by which the operation of smoking the rubber is more speedily performed than the mode at present in use. The little agricultural produce grown for home consumption is also in the hands of the Tapuyo, and is seldom undertaken by the white men.

Whilst nature yields so prodigally as in this country the abundant riches which man has but to put forth his hand to gather, without any real call on his energy or industry, no attempt will be made, no inducement is held out to an inert population to profit by the richness of the soil, or assist even that very nature by a little industry to increase their comfort and wealth.

Already the more accessible rubber districts are, it is said, becoming exhausted, and give a less yield than in former years; yet the rubber-bearing country is so vast, that the constantly newly discovered sources more than supply the deficiency occasioned by the exhaustion of the old, calls only for the slight exertion of extra travel, and has not yet imbued into this people the idea of planting the rubber-tree or caring for its growth.

From the foregoing remarks it will be seen that, as far as agricultural labour is concerned, little can be said beyond the repeated fact that, should proper encouragement be given by the offer of good land, and other early conveniences, to the gradual and voluntary emigration of an industrial race, who would not allow the extraction of rubber, or simple gathering of nature's produce to absorb their attention over all other labour, there is no doubt that in the course of a little time comfort and wealth would be enjoyed by the production of cotton, sugar, rice, cocoa, and other numerous articles indigenous to these provinces; and the condition of a country yet in the very infancy of civilization materially enhanced.

Food, or the necessary aliment of man, is cheap

and abundant; but luxuries are dear, being double and triple the price in England. The native labourer or workman requires little luxury: his meals are often procured and eaten near the site of his work, and consist generally of dried or salted fish, or meat with familia (flour of the mandioca root), which may not be so nourishing as bread, but is satisfying and pleasant to the palate; a cup of coffee, and a drink of cachass. There are no manufactories on any large scale in the provinces; cotton is neglected; and the small manufactories for bricks, soap, or the like, for home consumption, or again, workshops and workmen, are all yet in the very cradle of progress, and offer little assistance for forming correct statistics or ethics, as to class workmanship or bodies of workmen. Wages vary from 3 dollars to 5 dollars, and even 7 dollars mil reis per day, equal at an average exchange of 1s. 8d., to 5s., 8s. 4d., and 11s. 9d. sterling. The regular hours of labour are usually from six in the morning till four in the afternoon, from which time one hour, or one hour and a half, is exacted for a mid-day meal.

The purchase power of money in these provinces is, on the whole, immensely inferior to the standard in England. Luxuries are exorbitant, and only one or two of the necessaries are to be had cheaper than in England, or at what an English artisan would consider a reasonable rate. Thus, with reference to food, clothing, and house accommodation, it is found that in food only, the following articles compare favourably with English prices, viz. :—

Pará—Coffee, per lb., 10*d.* ; England, 1*s.* 8*d.*
    ,,   Beef   ,,   6*d.* ; England, 10*d.*

The quality of the beef, owing principally to the difficulties attending the transport of oxen from the interior, is, however, very poor, containing a large portion of bone, and on the whole the lesser price cannot be accepted as a fair criterion.

Bread, butter, flour, tea, and other table necessaries may be taken at double to triple English prices; fish, fowls, and other meats even exceed this proportion; and, notwithstanding the fertility of the soil, the purchase of vegetables is altogether beyond a working man's means, and can only be obtained by his own industry, should he be fortunate enough to secure a piece of ground for this purpose.

Household economy is in many instances difficult to practise, as it is found absolutely necessary to cook all fresh food when purchased, and nothing cooked can be kept in good condition for many hours.

House accommodation is difficult to obtain; rooms require to be large, and the necessity of ventilation prevents economy of space in building; so that, taking into account the high wages of artisans and labourers employed in their construction, the rent of buildings, which are of very inferior quality, is, on an average, three to four that of England. Thus, an artisan living in England in

to face page 298.

NEAR SANTAREM, PARA, BRAZIL

an £8 house, would not be more comfortable here at
a rent of £24 to £30; and even at these figures
would find considerable difficulty in obtaining a
house.

Clothing is of course lighter than in England,
and, with the exception of shoe-leather, should not
be more expensive; this article is, however, very
dear, as, whether from the humidity of the climate
or other causes, boots and shoes are quickly worn
out.

As may be concluded by these remarks, the
money power of £10 sterling in England may
be considered fully equal to £20 in Pará.

For the preservation of health in this climate,
temperance both in eating and drinking is certainly to be recommended; but not abstinence
either from meats, light wines, a moderate use of
beer, or a mixture with water of the native cachass,
as the exhaustive nature of the climate, producing a
continual perspiration, would otherwise weaken the
system.   On the other hand, over-indulgence, especially in drink,—a failing which, it is to be regretted,
many Englishmen are justly taxed with—quickly
produces bilious disorders, and often blood fevers,
resulting in many serious illnesses, and not a few
sudden deaths.

Over-exertion, exhaustion, or exposure to the
rain or sun, should be as far as possible avoided; at
the same time, a man need not be too careful of
himself, and a fair amount of work may be done
without danger: regular habits, and not too much

irritability of temper, are, of course, to be recommended.

Houses, as already mentioned, are difficult to be had, and have to be taken without always attending to convenience of locality.

Ventilation as a rule is good, drainage bad, and generally not over clean. In low-lying localities the houses are not free from miasma and air-poisonings; but such annoyances are not, on the whole, here to be dreaded; neither are the risks of health from these causes greater than in other parts of the world.

The standard of work is much lower than in England; but there is no doubt that in turning out work, high-classed artisans in Pará, as elsewhere, try to sustain their character. As "to making a stand against their turning out bad work,"—the quality of work being often, at the best, inferior, and the wages of workmen, or the charges made almost what such workmen choose to exact,—this question does not seem applicable to Pará, but is more so to places where the wholesale manufacture of class articles occurs. No doubt the sense of honour of many artisans would lead them to be careful in their work; but there is, I apprehend, no class of this kind.

To conclude, I have to observe that any intelligent, skilful, and sober artisan would be, and is, certain of remunerative employment in Pará. There are difficulties at first to be contended with, such as the language, house-accommodation, and occasional slight illnesses, until he becomes acclimatized and accustomed to the usages of the country. But a man who is persevering and not wasteful in his expenditure,—not too urgent for the pleasures of society,—of regular, quiet habits, avoiding the many vices that afflict strangers in Pará,—would, as a rule, be able considerably to better his circumstances, and in a brief period rise in reputation even above his class in England.

Unhappily, however, the greater portion, though not the whole of the English artisans who have hitherto appeared at Pará, seeking employment, have been sadly wanting in some of these qualities, especially in their extraordinary addiction to drink; and, in consequence, have but in a few cases realized their wishes, or reflected credit on their countrymen resident here, who feel ashamed of their daily intercourse with the more temperate Brazilians.

JAMES DE V. DRUMMOND HAY.

PARA, *September 16th*, 1870.

FINIS.

Printed in the United States
By Bookmasters